TD1050.S24 W39

Waxman, Michael F.,
1942-
Hazardous waste site
operations : a training
c1996.

0 1341 0075167 3

Date Due

MAR 23 1998			
DEC 1 4 1998			
MAY 2 6 1999			
NOV 2 2 1999			

NIAGARA COLLEGE

GLENDALE CAMPUS
LEARNING RESOURCE
CENTRE

HAZARDOUS WASTE SITE OPERATIONS

HAZARDOUS WASTE SITE OPERATIONS

A TRAINING MANUAL FOR SITE PROFESSIONALS

MICHAEL F. WAXMAN, Ph.D.
Associate Professor and Program Director
University of Wisconsin—Madison
Madison, Wisconsin

A WILEY–INTERSCIENCE PUBLICATION
John Wiley & Sons, Inc.
New York ■ Chichester ■ Brisbane ■ Toronto ■ Singapore

This text is printed on acid-free paper.

Copyright © 1996 by John Wiley & Sons, Inc.

All rights reserved. Published simultaneously in Canada.

Reproduction or translation of any part of this work beyond that permitted by Section 107 or 108 of the 1976 United States Copyright Act without the permission of the copyright owner is unlawful. Requests for permission or further information should be addressed to the Permissions Department, John Wiley & Sons, Inc., 605 Third Avenue, New York, NY 10158-0012.

This publication is designed to provide accurate and authoritative information in regard to the subject matter covered. It is sold with the understanding that the publisher is not engaged in rendering legal, accounting, or other professional services. If legal advice or other expert assistance is required, the services of a competent professional person should be sought.

Library of Congress Cataloging in Publication Data:

Waxman, Michael F., 1942–
    Hazardous waste site operations : a training manual for site professionals / by Michael F. Waxman
       p. cm.
    Includes index.
    ISBN 0-471-14218-2 (cloth : alk. paper)
    1. Hazardous waste sites—Safety measures. 2. Hazardous waste sites—Safety measures—Examinations, questions, etc. 3. Hazardous wastes—Law and legislation—United States. I. Title.
TD1050.S24W39   1996
628.4′2—dc20
                                                  95-52753

Printed in the United States of America

10 9 8 7 6 5 4 3 2 1

CONTENTS

PREFACE xi
ACKNOWLEDGMENTS xv

CHAPTER 1 ■ INTRODUCTION 1

CHAPTER 2 ■ OCCUPATIONAL HEALTH REGULATIONS, STANDARDS, AND GUIDELINES 7

  2.1 THE OCCUPATIONAL SAFETY AND HEALTH ACT 7
    2.1.1 Rights and Responsibilities of Employees Under the OSH Act 8
    2.1.2 Employer Responsibilities Under the OSH Act—More Details 13
  2.2 RESOURCE CONSERVATION AND RECOVERY ACT (RCRA) 14
    2.2.1 Introduction 14
    2.2.2 RCRA—Definition of a Hazardous Waste 14
  2.3 COMPREHENSIVE ENVIRONMENTAL RESPONSE, COMPENSATION, AND LIABILITY ACT (CERCLA) OF 1980 16
  2.4 SUPERFUND AMENDMENTS AND REAUTHORIZATION ACT (SARA) 16
    2.4.1 Hazardous Waste Operations and Emergency Response—"HAZWOPER" 17

2.5 DEPARTMENT OF TRANSPORTATION   20
    2.5.1 Regulatory History   20
    2.5.2 Hazardous Materials—Definitions   20
    REVIEW QUESTIONS   26

CHAPTER 3 ■ FUNDAMENTALS OF CHEMISTRY   29

3.1 BASIC CHEMICAL DEFINITIONS   30
3.2 THE PERIODIC TABLE   32
3.3 MOLECULES AND CHEMICAL BONDING   34
    3.3.1 Molecular Formula   34
    3.3.2 Ionic Bonding   35
    3.3.3 Covalent Bonding   36
3.4 OXIDATION–REDUCTION REACTIONS   39
    3.4.1 Oxidation State   40
3.5 ATOMIC AND MOLECULAR MASSES   41
    3.5.1 Atomic Mass   41
    3.5.2 Molecular Mass   42
    3.5.3 Molar Mass and Avogadro's Number   42
3.6 CHEMICAL REACTIONS AND BALANCING EQUATIONS   42
    3.6.1 Chemical Reaction   42
    3.6.2 Balancing Chemical Equations   43
3.7 SOLUTIONS   44
    3.7.1 Solution Concentration   45
    3.7.2 Electrolytes   47
3.8 ACIDS, BASES, AND NEUTRALIZATION   48
    3.8.1 Ionization of Water, Acids, and Bases   48
    3.8.2 Acid and Base Definitions   49
    3.8.3 Neutralization   51
3.9 TRACE INORGANIC CONTAMINANTS AND THE HEAVY METALS   53
3.10 ORGANIC CHEMISTRY   55
    3.10.1 Properties of Organic Compounds   55
    3.10.2 The Carbon Atom   56
    3.10.3 Aliphatic Hydrocarbons   58
    3.10.4 Aromatic Hydrocarbons   61
3.11 SUMMARY   65
    REVIEW QUESTIONS   65

CHAPTER 4 ■ INDUSTRIAL HYGIENE   69
  4.1 TOXICOLOGY   69
    4.1.1 Definitions   70
    4.1.2 Chemical Classifications   71
    4.1.3 Entry Routes of Toxicants into the Body   75
    4.1.4 Fate of Chemicals That Enter the Bloodstream   78
    4.1.5 Toxic Chemicals and Target Tissues   79
    4.1.6 Basis for Workplace Standards   82
    4.1.7 Threshold Limit Values (TLVs)   83
  4.2 CHEMICAL HAZARDS   88
    4.2.1 Hazard Communication Standard   88
    4.2.2 Chemical and Physical Characteristics   90
    4.2.3 Health Hazards   94
    4.2.4 Target Organ Effects   97
  4.3 PHYSICAL HAZARDS   107
    4.3.1 Fire   108
    4.3.2 Noise   110
    4.3.3 Temperature Stress   113
    4.3.4 Radiation   120
    REVIEW QUESTIONS   123

CHAPTER 5 ■ HAZARD RECOGNITION   125
  5.1 INTRODUCTION   126
  5.2 RECOGNIZING THE TYPES OF HAZARDS   126
    5.2.1 DOT—Labels and Placards   127
    5.2.2 The NFPA-704M System   130
    5.2.3 The HMIS (Hazardous Material Information System)   132
    5.2.4 Infectious Materials   134
    5.2.5 Recognizing Chemical Hazards   135
  5.3 CHEMICAL HAZARDS—OTHER RESOURCE MATERIALS   151
    5.3.1 Material Safety Data Sheets (MSDSs)   151
    REVIEW QUESTIONS   154

CHAPTER 6 ■ PERSONAL PROTECTIVE EQUIPMENT   163
  6.1 INTRODUCTION   163

6.2 REQUIREMENTS FOR THE USE OF PERSONAL PROTECTIVE EQUIPMENT   164
6.3 SELECTION OF PPE FOR SITE ENTRY   165
6.4 DEFINITIONS   166
6.5 SELECTION CRITERIA FOR DETERMINING THE PROPER CPC   168
    6.5.1 Summary: Selection Criteria for CPC   170
6.6 LEVELS OF PROTECTION—EPA DEFINITIONS   171
6.7 PRECAUTIONS WHEN WEARING CPC   177
    6.7.1 Permeation   178
6.8 INSPECTION, MAINTENANCE, AND STORAGE OF CPC   179
6.9 SUMMARY   180
    REVIEW QUESTIONS   182

CHAPTER 7 ■ RESPIRATORY PROTECTION   185

7.1 INTRODUCTION   185
7.2 RESPIRATORY PROTECTION—SELECTION CRITERIA   186
    7.2.1 Airborne Contaminants   186
    7.2.2 Respirator Selection   188
    7.2.3 Respirator Fit-Testing   189
7.3 RESPIRATORY EQUIPMENT   192
    7.3.1 Air-Purifying Respirators (APRs)   194
    7.3.2 Supplied-Air Respirators (SARs)   198
7.4 CARE OF RESPIRATORS   207
    7.4.1 Care of APRs   208
    7.4.2 Care of SCBAs   209
    7.4.3 Care of Airline Respirators   210
    7.4.4 Respirator Program   210
7.5 RESPIRATORY PROTECTION PROGRAM   211
    REVIEW QUESTIONS   224

CHAPTER 8 ■ MONITORING   227

8.1 INTRODUCTION   227
8.2 PURPOSE OF AIR MONITORING   228
    8.2.1 Understanding Concentration Measurements   228
8.3 MEASURING INSTRUMENTS   229
    8.3.1 Combustible and Oxygen-Deficient Atmospheres   232
    8.3.2 Toxic Atmospheres   237

8.3.3 Classroom Exercise: Direct-Reading Instruments   246
8.3.4 Direct-Reading Colorimetric Indicator (Detector) Tubes   260
8.3.5 Qualitative Chemical Identification Exercise Using Detector Tubes   266

8.4 PERSONAL AIR SAMPLING   271
8.4.1 Choosing Sampling Methods   271
8.4.2 Sampling Equipment and Procedures   278

8.5 SUMMARY   284
8.5.1 Types of Air Monitoring   286
8.5.2 Selecting Monitoring Equipment   287
REVIEW QUESTIONS   288

CHAPTER 9 ■ DECONTAMINATION   291
9.1 WORK ZONES   292
9.2 DECONTAMINATION PLAN   293
9.2.1 Decontamination Facilities Design   297
9.2.2 Decontamination Procedures   298
9.2.3 Recommended Equipment for the Decontamination   298

9.3 SUMMARY   310
REVIEW QUESTIONS   311

CHAPTER 10 ■ WORK PRACTICES   313
10.1 PLANNING AND ORGANIZATION   313
10.1.1 Initial Site Characterization and Analysis   314
10.1.2 Site Entry   315
10.1.3 Site Control   318

10.2 WORK PRACTICES   328
10.2.1 Confined Space Entry   328
10.2.2 Permit-Required Confined Space Entry Procedures   331
10.2.3 Emergency Response Planning   340
REVIEW QUESTIONS   349

CHAPTER 11 ■ DEVELOPING A SITE SAFETY PLAN   351
11.1 GENERAL REQUIREMENTS   352
11.2 SITE SAFETY PLAN   353
11.3 REVIEW QUESTIONS   367

CHAPTER 12 ■ MEDICAL SURVEILLANCE PROGRAM 369

- 12.1 THE MEDICAL EXAMINATION 370
  - 12.1.1 Types of Exams and When They Are Required 370
  - 12.1.2 The Examination Content 371
  - 12.1.3 Coverage and Payment for Medical Examinations 371
  - 12.1.4 Notification of Results 371
  - 12.1.5 Employee Access to Medical Records 372
  - 12.1.6 Employee Rights to Examination Records 373
- 12.2 TESTS AND PROTOCOLS 374
- 12.3 EXAMPLES OF HAZARDOUS CHEMICALS REQUIRING MEDICAL SURVEILLANCE 376
  - 12.3.1 Benzene 376
  - 12.3.2 Toluene and Xylene 379
  - 12.3.3 Formaldehyde (HCHO) 380
  - 12.3.4 Ketones 381
  - 12.3.5 Chlorinated Aliphatic Hydrocarbons 382
  - REVIEW QUESTIONS 385

CHAPTER 13 ■ RISK ASSESSMENT IN SUPERFUND SITE REMEDIATION 387

- 13.1 RISK ASSESSMENT AND RISK MANAGEMENT 388
- 13.2 DISCIPLINES INVOLVED IN THE PROCESS 388
- 13.3 THE FOUR FIELDS OF ANALYSIS 389
  - 13.3.1 Data Evaluation 389
  - 13.3.2 Toxicity Assessment 389
  - 13.3.3 Exposure Assessment 390
  - 13.3.4 Risk Characterization 391
- 13.4 RISK ASSESSMENT IN SUPERFUND 392
- 13.5 BASELINE RISK ASSESSMENT OF THE REMEDIAL INVESTIGATION 393
  - 13.5.1 Data Evaluation 396
  - 13.5.2 Exposure Assessment 396
  - 13.5.3 Toxicity Assessment 399
  - 13.5.4 Risk Characterization 402
- 13.6 EVALUATING REMEDIAL ALTERNATIVES 404
  - REVIEW QUESTIONS 405

GLOSSARY *407*

INDEX *433*

PREFACE

During the past 20 years, industry, government, and the general public have become increasingly aware of the environmental problems facing our nation. In 1962 *Silent Spring*, written by Rachel Carson, was published. Carson's book was one of the first that attracted national attention to the problems of toxic chemicals and the effects of these chemicals on the environment. *Silent Spring* recounted how the residues of the pesticide DDT could be found throughout the food chain. In aquatic birds, high levels of DDT were associated with reduced fertility. DDT affected the deposition of calcium in avian ovaries, leading to egg shells too thin to survive, thus causing a widespread reduction in many bird species.

In the late 1960s, an epidemic mercury poisoning occurred along Minamata Bay in Japan. Inorganic mercury from a chemical plant became methylated in the sediment of the bay and then bioaccumulated in shellfish. This shellfish was the major protein source for much of the local population. Ingestion of this contaminated shellfish caused hundreds of cases of paralysis and sensory loss.

In the late 1960s and mid-1970s, accidental contamination of rice cooking oil with polychlorinated biphenyls (PCBs) in Japan and Taiwan resulted in numerous miscarriages and birth defects. About the same time in the

United States, dairy cattle in Michigan became contaminated with polybrominated biphenyls (PBBs), causing widespread human exposures.

These events, the environmental contamination at Love Canal, New York, and Times Beach, Missouri, and the December 3, 1984 release of methyl isocyanate at a Union Carbide plant in Bhopal, India, which killed more than 2500 persons and seriously injured an estimated 150,000 more, generated intense media coverage and debate and eventually forced the Congress to act.

In 1980, Congress passed the Comprehensive Environmental Response, Compensation, and Liability Act (CERCLA), more commonly known as "Superfund." This Act sought to identify sites involving past releases of hazardous substances to the environment and to implement the remedial actions deemed necessary at each site to protect health and the environment. CERCLA and the Resource Conservation and Recovery Act of 1976, which regulated active sites of accumulation, treatment, storage, and disposal, were based on the legal concept that waste generators are liable for the long-term impact of their waste management practices, including their past practices.

After the passage of CERCLA, thousands of workers were hired to remediate the identified sites. No provisions were made in CERCLA for the training of this newly created work force. Many workers became ill or died because of this oversight. Many more were exposed to these high levels and did not develop symptoms, although they may develop chronic illnesses in the future.

On October 17, 1986, the Superfund Amendments and Reauthorization Act (SARA) was passed by Congress. This act specifically addressed the training requirements of site workers and supervisory personnel working on Superfund sites. From that date on, a new industry was founded. This industry, developed to meet the training needs of the hazardous waste remediation industries, is composed of all the training companies, institutions, colleges, and universities that formed divisions, organizations, or departments to accomplish the training requirements of SARA.

In the past decade, the backlog of training needs has been eliminated and new hires are being trained prior to the beginning of their site activities.

THE CHALLENGE

How does one go about teaching adults to work safely? Obviously, much of the answer lies in the experience and the willingness of the trainee and the expertise of the trainer.

The challenge to the educator or trainer is to determine how the experience of health and safety experts can be translated or transformed to the educational or training setting and, then, how to transfer this experience to the students so that they gain the insight of the experience of others. This information must be in a usable form so that trainees can return to work and use it immediately while performing their daily duties.

Since 1986, the University of Wisconsin and numerous other training institutions have developed syllabi, with the help of numerous remediation experts, to meet this challenge.

Workers are now fully trained about the current federal and state regulations that govern their activities on-site. They receive information about the health risks associated with the chemicals on their work site, and they learn how to use personal protective equipment and how to follow safe work practices. Therefore, current workers are far more capable of handling their duties safely than their predecessors.

The term "worker" as used in this book refers to anyone—engineer, hydrogeologist, technician, heavy equipment operator, or laborer—who, in the normal course of his or her job activities, may be exposed to hazardous or toxic substances.

This text was written and designed as a hazardous waste site operations training manual/text to meet the needs of both trainers and trainees. As a program director and trainer at one of the original organizations providing the OSHA-required 40-hour Hazardous Waste Site Operations training, the Department of Engineering Professional Development, College of Engineering at the University of Wisconsin-Madison, I found that appropriate training materials were often unavailable or inadequate. In most cases, training manuals were providing an outline or skeleton of the training curriculum; each trainer then had to "flesh out" his or her section(s) of the outline. In many cases, the trainers added too little content, so the trainees had to take copious notes. But exhaustive note taking can create a serious problem; for instance, the OSHA 40-hour training programs are very intensive, usually covering extensive materials in a very short period of time, while incorporating considerable hands-on training. Most students cannot keep up with the trainers presentations while also taking detailed notes. Therefore, students were leaving the training course with incomplete notes and materials that could serve only as a limited reference source during their subsequent work activities.

This text incorporates much of the heretofore lacking information that trainees and trainers have needed. The book also addresses the numerous regulations, which are continuously changing in both content and interpretation.

This text, then, provides a complete training manual for both the instructor to use as a guide and the trainee or student to use as a reference. Use of this text will produce both uniformity and completeness in the materials and the training of site workers.

Writing this book is my response to the challenge just described. By enabling site workers to perform their jobs safely and by giving trainers a practical tool with which they can instruct and guide, I hope to ensure their health and safety and, indeed, the health and safety of our entire population.

Michael F. Waxman
Madison, Wisconsin

ACKNOWLEDGMENTS

Writing a textbook that required more than two years to come to fruition obviously involved many people along the way whose contributions should be acknowledged. Many people have provided a significant contribution to this book; a few I can name, but to many I express sincere appreciation.

Perhaps the most important people who provided insight and impetus to this project were the thousands of students in the many training sessions the university held in the past decade. These students acted essentially as "test subjects" in our numerous hands-on and table-top exercises. Their performance and critical comments helped to make our training exercises as beneficial as possible. Through classroom and field practice, we developed many presentations, demonstrations, and exercises that could be completed in the allotted 40 hours. Without the help of our students, the success of our program could not have been achieved.

The trainers are another group deserving special recognition. These professionals utilized not only their skills as health and safety experts, but also their talents as trainers and educators; they prepared their materials enthusiastically to achieve the necessary goals of the training sessions. Our trainers also provided much of the expertise in this book. Over these past years, their expertise has enriched the lives of their students and fellow trainers—and my life as well. I feel that I have benefited as much as the students. These benefits are reflected in this text.

I wish to thank several expert reviewers who helped with chapters of the manuscript. These include Dr. John T. Quigley and Mr. John Konig, University of Wisconsin, and many others too numerous to mention.

Another person I must not forget to acknowledge for his expert assistance is Mr. Darrell Petska, Department of Engineering Professional Development, University of Wisconsin. His numerous editorial comments were invaluable to the completion of this manuscript, and I appreciate his talented input during the past several years.

M.F.W.

INTRODUCTION

A new industry has been created as a response to the needs of the hazardous waste industry. This industry is mainly concerned with the health and safety training of workers and the transfer of regulatory information to those persons engaged in hazardous waste operations. This manual and the training programs that utilize it and the many other resources available have been developed to help train the thousands of workers engaged at these sites.

The hazardous waste remediation industry has also evolved from the promulgation of regulations and their enforcement by the Environmental Protection Agency (EPA). The EPA requires site assessments and remediation to remove the hazards from suspected and environmentally unsound waste sites. Because of this country's need for a large work force, workers have often been employed with little or no experience in handling toxic chemicals or hazardous wastes. Regulations clearly needed to be developed to provide for worker safety and health at these sites.

In 1986, Congress reauthorized the modified Comprehensive Environmental Resource Compensation and Liability Act of 1980, known as "Superfund." This modified Superfund Act, called the Superfund Amendments and Reauthorization Act of 1986 (SARA), detailed the training requirements for persons engaged in hazardous waste operations and emergency response. This text has been developed to help the hazardous waste remediation industry (and generators of hazardous wastes) to train their personnel and to comply with the aspects of these standards.

2 ■ INTRODUCTION

Hazardous waste sites are dangerous because they are unpredictable: No one knows for sure what is buried in an abandoned site until the site assessment is completed. And of course, surprises can still occur, even after knowing what is buried at a site. Also, site characteristics change as work progresses. Heavy equipment used for the removal of contaminated soil and the excavation of buried drums can create additional hazards. The excavation of trenches and pits create potential confined space hazards and a host of other physical hazards. The bulk packing or containerization of compatible wastes creates another set of different hazards.

The industrial workplace also poses many health and safety risks. The hazards associated with the workplace can be specific to the manufacturing process, the raw materials used in that process, and the by-products or wastes generated by that process. In addition, many industrial facilities are also associated with remediation efforts.

The hazards found at both the hazardous waste site and the industrial workplace can include

- Chemical exposure
- Fire and explosion
- Oxygen deficiency
- Ionizing radiation
- Biological hazards
- Physical safety hazards
- Electrical hazards
- Heat stress
- Cold exposure
- Noise

The interaction and possible synergistic action among the many chemical substances used or produced in the workplace may pose additional hazards not originally present.

Furthermore, workers engaged in site assessment, sampling, and remediation are subject to dangers posed by the disorderly physical environment of these uncontrolled sites. And the stress of working in protective clothing adds its own risks.

This text emphasizes the need for a companywide health and safety program, which must be developed to provide workers with the necessary information and training to protect against potential hazards and specifically known hazards. The health and safety program must be continuously updated with new information and details on changing site conditions. By knowing the specifics of the plan and using a checklist for the site, companies can maintain control over hazard exposure during the actual work periods.

This text addresses the major training requirements of OSHA Standard 29 CFR 1910.120; it also provides the reader, student, and/or trainer with additional information and exercises that will help workers and their companies meet all the training requirements.

Chapter 2 reviews the pertinent OSHA regulations governing the rights and responsibilities of the employer and employee at remediation site cleanups and emergency response actions. This section also describes several regulations promulgated by the EPA and the Department of Transportation (DOT) that impact work at these sites.

Chapter 3 is written to provide the trainee having little or no background in chemistry with the basics of environmental chemistry. These fundamentals are designed to be simple, understandable, and interesting and provide site professionals with some of the chemical information they should know about the materials that may be encountered on the site.

Chapter 4 covers a major concern of workers who handle hazardous materials: how these materials enter and interact with their bodies. The chapter then focuses on the major sources of injury at these workplaces—that is, the physical hazards of the site and the methods to reduce them.

Hazard recognition, Chapter 5, is crucial in the prevention of worker exposure. Workers must understand the compatibility of chemicals, labeling and placarding requirements, and methods of handling and sampling drums. They must also know how to read material safety data sheets (MSDS). Such knowledge will assist in the recognition of site hazards and prevent needless exposures.

Two of the most important chapters in this text are Chapters 6 and 7; these chapters discuss personal protective equipment (PPE) and respiratory protection, respectively. These sections detail protective clothing and respiratory equipment from head to toe and relate respiratory protection and PPE with environments by degree of hazard.

The ability to measure the air quality of the environment and workplace is crucial in determining the levels of respiratory protection and PPE required to protect the health and safety of the site workers. Chapter 8 describes the equipment and strategies that must be considered prior to site entry and subsequent activity.

Decontamination procedures for personnel and equipment are described in Chapter 9. This chapter discusses the equipment needed for decon and practical methods to determine the adequacy of the procedures. It also presents examples of decon plans to illustrate the concepts presented.

Safe work practices (Chapter 10) will instruct the reader on the proper methods and procedures for site characterization and analysis, site control, confined space entry and planning for site emergencies. A knowledge and understanding of these work practices are a must for all site personnel.

Chapter 11 ("Developing a Site Safety Plan") provides the reader with a generic site safety plan that can be used to help in the development of a specific plan. By knowing the specifics of the plan and familiarizing themselves with the site map, workers can maintain a level of control over the hazards of the site and minimize potential exposures.

Chapter 12 ("Medical Surveillance Programs") discusses the OSHA requirements for determining worker health and possible exposures through medical testing. The chapter also provides information on interpreting the results.

Because risk assessment is becoming the cornerstone of our environmental policy-making, a general discussion of risk assessment has been included as Chapter 13.

This text is a comprehensive guide for worker safety during remediation site cleanups and response actions. However, companies or workers should have at hand other references to complement it and provide additional information that is not in the scope of this text.

This text will serve as a resource and guide to be used often by those workers whose profession necessitates their interaction with hazardous wastes. This text, along with the proper training, will ensure that the quality of life of site workers will not be diminished by their job activities.

REFERENCES

Comprehensive Environmental Response, Compensation and Liability Act (Superfund), PL 96-510, 42 USC section 9601 et seq., 1980.

REFERENCES ■ 5

Levine, S., and W. Martin, eds., *Protecting Personnel at Hazardous Waste Sites*, Butterworth, Stoneham, MA, 1985.

OSHA 29 CFR 1910.120: *Hazardous Waste Operations and Emergency Response*, Final Rule, U.S. Government Printing Office, Washington, D.C., 1989.

Superfund Amendments and Reauthorization Act, P.L. 99-499, October 17, 1986.

CHAPTER 2

OCCUPATIONAL HEALTH REGULATIONS, STANDARDS, AND GUIDELINES

CHAPTER OBJECTIVES

When you have completed this chapter, you will be better able to

- Identify the relevant government agencies and some health and safety regulations that they enforce
- Identify federal agencies and the environmental regulations they enforce
- Describe key rights and responsibilities that workers and employers have under the OSH Act

2.1 THE OCCUPATIONAL SAFETY AND HEALTH ACT

To ensure worker protection at the job site, Congress passed the Occupational Safety and Health Act (OSH Act), which was signed into law on December 29, 1970. The OSH Act directed the secretary of labor to establish the Occupational Health and Safety Administration and authorized the creation of the office of the Assistant Secretary of Labor. That office was to direct OSHA to

- Promulgate standards for worker safety and health
- Enter and inspect workplaces
- Issue citations for violations of standards and recommend penalties
- Oversee state plans for safety and health
- Encourage the development of training programs for workers, management, and health professionals

The OSH Act serves to ensure safe and healthful conditions in the workplace. The act requires that each employer furnish a workplace ". . . free from recognized hazards that are causing or are likely to cause death or serious physical harm to employees."

This statement, known as the general duty clause, is a comprehensive provision that can be used to cite employers for serious situations for which no standard exists. The act also requires that employees comply with all applicable rules, regulations and standards pertaining to occupational safety and health.

In addition to the establishment of OSHA, the National Institute for Occupational Safety and Health (NIOSH) was created as a research agency within the Department of Health and Human Services. NIOSH conducts research on occupational hazards and makes recommendations to OSHA regarding the creation or revision of OSHA standards. NIOSH has the responsibility to evaluate personal protective equipment and hazard-measuring instruments, such as those commonly used at hazardous waste sites and in industrial facilities. NIOSH also provides education and training in occupational safety and health.

OSHA standards are legally enforceable sets of industry-specific regulations intended to address concerns for safety and health of workers. These standards are developed and revised constantly. Employers, employees, and all other interested parties have the opportunity to comment on proposed new standards or revisions.

2.1.1 Rights and Responsibilities of Employees Under the OSH Act

The OSH Act gave certain rights to the American worker. Most significantly, workers have the general right to a safe and healthy work environment. The act also provides for specific worker rights (OSHA, 1991).

Under the OSH Act, workers have the right to

- File an OSHA complaint if unsafe conditions exist in the workplace
- Obtain records of workplace hazards, including annual summaries of recordable injuries and illness
- Have access to information about hazardous chemicals used in the workplace
- Request an OSHA inspection of the workplace
- Participate in an OSHA walk-around inspection
- Give information or speak with an OSHA compliance officer

Some worker responsibilities under the OSH Act are to

- Follow reasonable employer safety rules
- Wear or use required safety equipment
- Seek medical treatment promptly when injured
- Bring safety and health concerns to the attention of the supervisor or management

Some employer responsibilities under OSH Act are to

- Provide a safe and healthy workplace
- Comply with OSHA standards
- Maintain appropriate records of injuries and post these at least annually

Worker Rights—More Details The OSHA law provides workers with a number of important rights that have been discussed in general terms above. The following is a closer look at some of the more significant rights as they relate to workers.

To File a Complaint: If a violation is believed to exist, workers or their authorized representatives have the right to file a complaint with OSHA and to remain anonymous to the employer, if desired. The complaint should describe the location of the employer, as well as the nature and location of the hazard.

To Have an Inspection of a Workplace: Once a complaint is received at the OSHA area office, it will be assigned to one of their compliance officers, according to a priority that is defined by law. The inspection priority is

1. Imminent danger
2. Catastrophic accident (a fatality, or five or more workers hospitalized overnight as a result of on-the-job exposure)
3. Complaint inspection (as a result of a worker or union request)
4. Scheduled inspection (general OSHA inspection not because of a complaint or catastrophe but because injury statistics show that the employer has more injuries and illnesses than similar employers)

To Participate in the OSHA Walk-Around Inspection: Through an employee organization, such as a union, an employee representative is designated to accompany the OSHA compliance officer in the walk-around inspection of the establishment. The union is advised to notify in writing both the company and OSHA of their selection for walk-around representatives. OSHA regulations currently do not require the employer to pay the employee for time spent on the OSHA walk-around. Walk-around activities include all opening and closing conferences related to the conduct of the inspection but do not include any post-citation appeal procedures.

To Be a Witness or to Give Information: Every employee has the right to appear as a witness at an OSHA hearing. During the walk-around inspection, or before or after the inspection, any employee has the right to provide OSHA with any information regarding possible safety and health hazards.

To Be Informed of Imminent Dangers: All employees have the right to be informed by the OSHA compliance officer if they are exposed to an imminent danger (one which could cause death or serious injury now or in the near future). The compliance officer will also ask the employer to voluntarily stop the particular work process and remove the employees. A judge can force the employer to do this if the employer refuses to comply with the officer's, request.

To Be Told About Citations: Notices of OSHA citations must be posted in the workplace near the place where the violation occurred. Notices must remain posted for three days or until the hazard is corrected, whichever is longer. Citations and penalty notification forms are generally available upon request from the OSHA area office. When an OSHA industrial hygiene inspection has taken place, the hygienist's report, which includes substances collected, procedure used, and measurement results, may also be obtained upon request by the employees, their representatives, or their union.

To Appeal About OSHA Performance: If OSHA fails to perform in a responsible and timely manner, the employees, employer, or the union have the right to meet with the OSHA area director or the OSHA regional administrator and ultimately appeal to the secretary of labor.

To Appeal Abatement Dates (When a Violation Must Be Corrected): The findings of the OSHA officer may be appealed within 15 working days of the issuance of the citation to the employer. An employee's or union's right to contest citation is limited only to the question of the reasonableness of the abatement period of the citation. Employees or their union cannot contest the penalty amount or the citation itself.

To a Closing Conference After an Inspection: Employees have the right to meet privately with the OSHA officer and discuss the results of the inspection. OSHA procedures state that the OSHA inspector shall inform the employers and employees that a generally responsive discussion covering general issues will be held.

To Know of Health Hazard Exposures: Employees have the right to be notified if exposed to occupational health hazards and to be notified of the results of occupational health studies conducted by the employer or OSHA officers. The employees or their union can and should ask for any and all instrument readings or levels of contaminants found. A copy of the lab report should also be requested from OSHA. These documents are normally available upon request but may also be obtained by any member of the public pursuant to the Freedom of Information Act.

To Have Access to OSHA Records: Generally speaking, most OSHA records are available upon request (after the initial 15-day appeal period unless the case is appealed). The union or employees should contact the OSHA area office where the plant is located.

To Participate in Development of New Standards: Every employee has the right to participate in the development of new safety and health standards or modification of old codes. Individuals may also comment on proposed standards during open discussion periods.

To Review a Citation Procedure When a Citation Is Not Issued: Every employee has the right to request an informal review of an OSHA officer's refusal to issue a citation, or any other issue related to an inspection, citation, notice of proposed penalty, or notice of intention to contest a citation. A written statement as to why a citation was not issued in particular instances may be requested.

To File a Discrimination Complaint: If an employee has been discriminated against as a result of exercising his or her rights under OSHA, he or she has the right to file a complaint with the OSHA area office within 30 days. This time limit is very strictly enforced. Similar right to file a complaint may exist with state and local anti-discrimination agencies, as well as the union in organized workplaces.

Worker Responsibilities—More Details

To Abide by Established Safety Rules: Workers cannot be cited or fined by OSHA, but employers can take disciplinary action for violation of established safety rules.

To Follow All OSHA Regulations: Workers are normally required to follow reasonable workplace safety rules established by the employer. Workers are also normally required to follow all OSHA regulations.

To Wear and/or Use Required Safety Equipment: Workers are responsible for wearing and/or using required safety equipment. This is subject to agreement between the employees and employer, however. An example is hearing protection. The employer is required to furnish hearing protection devices to workers and to maintain them, but many employees take care of their own hearing protection. The same is often true for respiratory protection. Safety footwear, in many industries, is furnished by employees, even though OSHA standards may require that employers provide it. In some cases a footwear allowance is available.

To Seek Prompt Medical Treatment When Required: Workers should promptly seek medical treatment when required. Depending on applicable state law, workers have a right to be treated by a physician of their own choice for work-related injuries. The key here is not to delay medical treatment when necessary.

To Bring Safety and Health Concerns to the Attention of Management. Workers should bring safety and health hazards or concerns to the attention of their supervisors as soon as possible. If the worker is a member of a union, he/she may want to ask the union representative to bring the issue to the attention of management.

2.1.2 Employer Responsibilities Under the OSH Act—More Details

The OSH Act places important responsibilities on the employer. An understanding of these responsibilities is essential to maintaining a safe and healthy workplace and meeting other requirements of the law.

To Furnish a Safe and Healthy Job and Work Environment: The employer must furnish each "employee employment and a place of employment which are free from recognized hazards that are causing or are likely to cause death or serious physical harm" to employees. This is commonly referred to as the general duty clause of the Act. It describes the overall or general responsibility of the employer to protect employees from harmful situations or chemicals.

To Comply with OSHA Standards: Employers must comply with both the OSHA General Industry Standard (1910), including the HAZWOPER Standard (1910.120). HAZWOPER applies only to hazardous waste operations and related emergency response. In the event of conflict between HAZWOPER and the general standard, the most protective must be enforced. The General Industry Standard covers virtually all production industries. It has some important requirements, which include the following:

1910.20	Employee access to exposure and medical records
1910.38	Employer emergency plans and fire prevention plans
1910.134	Respiratory protection
1910.156	Fire brigades
1910.1000	Air contaminants
1910.1200	The right-to-know (hazard communication) standard

To Maintain Records of Injuries. Under the OSH Act, the employer must maintain a log of injuries and make it available to OSHA compliance officers upon request. Each year (by no later than February 1) the employer must post an annual summary of the injury log for the information of the employees. This form is called the OSHA 200. The employer must also display the required OSHA poster, which outlines specifics of the OSH Act.

2.2 RESOURCE CONSERVATION AND RECOVERY ACT (RCRA)

2.2.1 Introduction

The Resource Conservation and Recovery Act (RCRA) was enacted as Public Law 94-580 in 1976 as an amendment to the Solid Waste Disposal Act (SWDA). The primary objective of RCRA is to protect human health and the environment. A secondary objective is to conserve valuable material and energy resources by (a) providing assistance to state and local governments for prohibiting open dumping, (b) regulating the management of hazardous wastes, (c) encouraging recycling, reuse, and treatment of hazardous wastes, and (d) providing guidelines for solid waste management and resource recovery. RCRA provides for "cradle to grave" tracking of hazardous wastes, from the generator to transporter to treatment, storage, or disposal.

RCRA's major provisions for controlling hazardous waste include

- Standards for generators and transporters of hazardous wastes
- A manifest system for tracking hazardous wastes from generation to final disposal
- Permit requirements for facilities that treat, store, or dispose of hazardous waste (TSD facilities)
- Requirements for state-administered hazardous waste programs

2.2.2 RCRA—Definition of a Hazardous Waste

RCRA defines a hazardous waste as a waste that may cause substantial damage to the health or environment when improperly managed. The hazardous wastes regulated under RCRA are either specified on a list of specific wastes or possess any of the following four waste characteristics:

Ignitability: The characteristic of ignitability was established by the EPA to identify wastes capable of causing a fire or accelerating a fire during routine waste handling. A waste exhibits the characteristics of ignitability if it meets one or more of the following parameters:

- It is a liquid with a flash point of less than 140°F.
- It is a nonliquid that under normal conditions may cause a fire by absorption of moisture, by friction, or by spontaneous chemical reaction; such nonliquid burns so vigorously when ignited that it creates a hazard.
- It is an ignitable compressed gas by DOT regulations provided in 49 CFR 173.300.
- It is an oxidizer by DOT definition provided in 49 CFR 173.151.

Corrosivity: The characteristic of corrosivity was established to regulate wastes capable of corroding metal containers and potentially releasing other wastes. The characteristic of corrosivity is also assigned to wastes with a pH of less than or equal to 2.0 or greater than or equal to 12.5, or a liquid that corrodes steel at a rate greater than 0.25 inches per year under a specific set of conditions.

Reactivity: The waste characteristic of reactivity was established by the EPA to regulate extremely unstable wastes and those having a tendency to react violently or explode when the waste is being mismanaged. The RCRA regulations do not provide precise scientific descriptions of test protocols for measuring reactivity because these protocols are unavailable. Instead, RCRA provides a descriptive definition of reactivity (such as behavior of the substance when mixed with water, when heated, when mixed with a fuel, etc.).

Toxicity Characteristic (TC) and Toxicity Characteristic Leaching Procedure (TCLP): The TC toxicity characteristic was established to identify leachable hazardous concentrations of specific toxic heavy metals or pesticides capable of migrating into the groundwater under conditions occurring in a landfill. The EPA has established a toxicity characteristic testing protocol in which the liquid extract from the procedure is analyzed; if the level of waste constitutent exceeds regulatory levels, the waste is considered hazardous.

2.3 COMPREHENSIVE ENVIRONMENTAL RESPONSE, COMPENSATION, AND LIABILITY ACT (CERCLA) OF 1980

The CERCLA Act, popularly called "Superfund," was enacted for emergencies and cleanup at uncontrolled waste sites. It was the first federal law to respond to releases of hazardous substances or other contaminants into the air, water, or land. The act provides federal funding to respond to the releases; however, the parties who are responsible for the release must perform the cleanup work. If the responsible parties do not take appropriate action, the EPA has the authority to order them to do so. If they still refuse to respond, the EPA may use federal funding and seek recovery from the responsible parties. This law is especially important to local hazardous materials response (HAZMAT) teams because it provides a legal route of reimbursement to the team for materials and supplies expended when responding to a hazardous substance release.

According to sections 103A and B of CERCLA, the National Response Center must receive immediate notification from persons in charge of facilities from which hazardous substances have been released in quantities that are equal to or greater than the reportable quantities (RQs) of designated chemicals in the CERCLA regulation.

2.4 SUPERFUND AMENDMENTS AND REAUTHORIZATION ACT (SARA)

In October 1986, Congress passed the five-year reauthorization of Superfund, entitled "Superfund Amendments and Reauthorization Act" (SARA). The intent of Congress was to better safeguard the health and safety of workers and the community at large. SARA consists of three separate sections or "titles." Titles I and III deal with training, emergency response and planning, while Title II concerns a fund for hazardous waste cleanup. In brief, the titles require the following:

Title I: Training of emergency response personnel and workers at hazardous waste operation sites; preparation of a written inventories and emergency response plan for companies where hazardous materials are stored and may be spilled or released; proper procedures for handling emergency response operations, including the use of an incident command system (ICS).

2.4 SUPERFUND AMENDMENTS AND REAUTHORIZATION ACT (SARA)

Title II: "Superfund" authority to pay for hazardous waste cleanup through a tax on industry and EPA authority to seek recovery from the responsible parties.

Title III (Community Right-to-Know): Development of comprehensive community emergency plans by Local Emergency Planning Committees (LEPCs); reporting of certain chemical inventory and release information to fire departments, LEPCs, and state emergency response commissions (SERCs).

2.4.1 Hazardous Waste Operations and Emergency Response—"HAZWOPER"

SARA Title I required the Occupational Safety and Health Administration (OSHA) to develop a standard to protect hazardous waste site workers and emergency response personnel. That standard is commonly called HAZWOPER or 29 CFR 1910.120.

"Section 1910.120" refers to the written standard's location in federal regulation documents. This number reflects the standardized system the federal government uses to index all of its regulations.

In the full citation for HAZWOPER, 29 CFR 1910.120, the 29 indicates OSHA regulations that are located in Title 29. The abbreviation CFR indicates the Code of Federal Regulations, which is the title of the government publication. The part number 1910 covers general industry, and 120 is the section number covering hazardous waste operations and emergency response.

More letters and numbers after the section identify the exact paragraph. A more complete reference for site worker training is (e)(1). The course, Health and Safety Training for Hazardous Waste Operations, which this text presents, is designed to meet the initial training requirements for personnel at this level.

Training Requirements of HAZWOPER This standard applies to workers who clean up uncontrolled hazardous waste sites and RCRA remediation sites; underground storage tank workers performing remediation activities; emergency response workers such as firefighters and police who respond to hazardous materials spills; and workers at waste treatment, storage, and disposal facilities (TSDFs).

Requirements for general site workers are as follows:

- Understand hazardous materials and associated risks.
- Understand the safety and health program and read the site safety plan prior to working at that site.
- Understand the elements of site characterization and analysis.
- Understand the elements of site control.
- Be aware of the medical surveillance program and its requirements.
- Understand the engineering controls, work practices, and use of personal protective equipment for worker protection.
- Select and use proper personal protective equipment that is provided.
- Understand the use of air monitoring equipment and be familiar with the different types of monitoring equipment and their uses and limitations.
- Be familiar with proper drum and container handling methods.
- Understand proper decontamination procedures.
- Understand the elements of the site emergency response program.

Requirements for *Emergency Response—Awareness Level* are as follows:

- Understand hazardous materials and associated risks.
- Understand potential outcomes of emergencies.
- Be able to recognize the hazards.
- Identify hazardous materials.
- Understand role of emergency responder.
- Be able to contact appropriate personnel.

Emergency responders—operations level are individuals who respond to releases or potential releases of hazardous substances as part of the initial response for the purpose of protecting nearby persons, property, or the environment from the effects of the release. They are trained to respond in a defensive fashion without actually trying to stop the release. Their function is to contain the release from a safe distance, keep it from spreading, and prevent exposures. First responders at the operational level must have received at least eight hours of training or have sufficient experience to objectively demonstrate competency in the following areas and in those listed for the awareness level.

Requirements for the *Emergency Response—Operations Level* are as follows:

2.4 SUPERFUND AMENDMENTS AND REAUTHORIZATION ACT (SARA) ■ 19

- Fulfill requirements of awareness level.
- Have knowledge of the basic hazard and risk assessment techniques.
- Know how to select and use proper personal protective equipment.
- Understand the basic hazardous materials terms.
- Know how to perform the basic control, containment, and confinement operations.
- Understand and know how to implement the basic decontamination procedures.
- Understand the relevant standard operating procedures and termination procedures.

Hazardous materials technicians are individuals who respond to releases or potential releases for the purpose of stopping the release. They assume a more aggressive role than the first responder at the operations level in that they will approach the point of release of a hazardous substance. Hazardous materials technicians must receive at least 24 hours of training including awareness and the first responder operations level. They must also have competency in the following areas:

- Know how to implement the emergency response plan.
- Know the classification, identification, and verification of known and unknown materials by using field survey instruments and equipment.
- Be able to function within an assigned role in the incident command system.
- Know how to select the proper personal protective equipment.
- Understand hazard and risk assessment techniques.
- Be able to perform advanced control, containment, and/or confinement operations.
- Understand and implement decontamination procedures.
- Understand termination procedures.
- Understand basic chemical and toxicological terminology and behavior.

There are *three distinct levels of training*; they should not be confused or substituted:

1. Awareness
2. Operations level, first responder
3. Hazardous materials technician

2.5 DEPARTMENT OF TRANSPORTATION

2.5.1 Regulatory History

There have been numerous laws passed by congress since the 1800s. Many of these regulated the transportation of chemicals in interstate commerce.

The Department of Transportation Act of 1966 This law was enacted on October 16, 1966. This new law authorized the establishment of the Department of Transportation, effective on April 1, 1967, and transferred to the new Department the regulatory control and responsibility previously vested with the ICC, the Federal Aviation Agency, and the Coast Guard.

The Transportation Safety Act of 1974 This law went into effect on January 4, 1976. It centralized in the Secretary of Transportation the authority to promulgate and enforce hazardous materials regulations for all modes of transportation. These regulations may govern any safety aspect of transporting hazardous materials, including the packaging, repacking, handling, labeling, marking, placarding, and routing.

This law also authorizes the Secretary to establish criteria for handling hazardous materials. These criteria included the establishment of qualifications and training levels for personnel involved in the transportation of hazardous materials shipments.

HM-181 This regulation went into effect on January 1, 1991. This regulation is based on the United Nations (U.N.) Recommendations on the Transportation of Dangerous Goods. New labels and placards, shipping containers, and shipping descriptions were mandated. The U.N. system simplifies and reduces the number of regulations. Twenty POP (performance-oriented packaging) standards have replaced 100 HMR (hazardous materials regulations) specifications. These changes will facilitate international commerce. A package can be sent to Seattle or Singapore with the same labeling system.

2.5.2 Hazardous Materials—Definitions

The shipper must determine whether the materials to be transported have the characteristics of one or more of the hazardous materials defined in the hazardous materials regulations of the U.S. Department of Transpor-

2.5 DEPARTMENT OF TRANSPORTATION ■ 21

tation. The shipper using the Hazardous Materials Table (Sec. 172.101) can determine whether or not the materials to be shipped are regulated as hazardous materials. However, it must be emphasized that a material may not be listed by name in the table and still be subject to regulation. Furthermore, some of the materials listed in the table may not be subject to these regulations due to the quantity to be transported, the transportation mode to be used, or a number of other reasons.

The Transportation Act of 1974 defined a hazardous material as

a substance or material in a quantity and form which may pose an unreasonable risk to health and safety or property when transported in commerce.

Under the provisions of the Transportation Act of 1974, the Secretary of Transportation was granted the authority to designate what is a hazardous material. These materials may include, but are not limited to, the following: explosive, radioactive materials; etiologic agents; flammable liquids or solids; combustible liquids or solids; poisons; oxidizing or corrosive materials; and compressed gases.

Title 49 CFR contains specific definitions of the various types of hazardous materials subject to the regulations of the U.S. DOT issued under the authority of the Act of 1974. Within these rules, the following hazardous materials are defined (see HM-181):

Class 1 (explosive) materials
    1.1 Mass explosive hazard
    1.2 Projection hazard
    1.3 Mass fire hazard
    1.4 Minor explosion hazard
    1.5 Very insensitive hazard
    1.6 Extremely insensitive hazard

Class 2 (gases) materials
    2.1 Flammable gas
    2.2 Nonflammable gas
    2.3 Poisonous (toxic gas)

Class 3 (flammable liquid) materials

Class 4 (flammable solid) materials
    4.1 Flammable solid
    4.2 Spontaneously combustible
    4.3 Dangerous when wet

Class 5 (oxidizer/peroxide) materials
    5.1 Oxidizer
    5.2 Organic peroxide
Class 6 (poison) materials
    6.1 Poisonous (toxic)
    6.2 Infectious substance
Class 7 (radioactive) materials
Class 8 (corrosive) materials
Class 9 (miscellaneous) materials include hazardous wastes
Other regulated materials (ORM-D)

The hazardous materials definitions explained on the following pages are derived from the appropriate sections within the hazardous materials regulation. These definitions may be used as general guidelines to defining material to be shipped, but the specific regulations should be consulted for the full specifications of each material considered a hazardous material.

Class 1 (Explosive) Materials An explosive is any substance or article, including a device which is designed to function by explosion or which, by chemical reaction within itself, is able to function in a similar manner even if not designed to function by explosion, unless the substance or article is otherwise classed. Class 1 is divided into six divisions as follows:

Division 1.1 consists of explosives that have a mass explosion hazard. A mass explosion is one which affects almost the entire load simultaneously.

Division 1.2 consists of explosives that have a projection hazard but not a mass explosion (e.g., grenades, handgun, or rifle).

Division 1.3 consists of explosives that have a fire hazard and either a minor blast hazard or a minor projection hazard or both, but not a mass explosion hazard.

Division 1.4 consists of explosive devices that present a minor explosion hazard. No device in this division may contain more than 25 g (0.9 ounces) of a detonating material.

Division 1.5 consists of very insensitive explosives. This division is comprised of substances which have a mass explosion hazard but are so insensitive that there is very little probability of initiation or transition from burning to detonation under normal conditions of transport.

Division 1.6 consists of extremely insensitive articles which do not have a mass explosive hazard. This division is comprised of articles which contain only extremely insensitive detonating substances and which demonstrate a negligible probably of accidental initiation or propagation.

Class 2 (Gases) Materials Class 2 is divided into three divisions as follows:

Division 2.1 materials are those which are a gas at 20°C (68°F) or less and 101 kPa (14.7 psi) pressure and which (a) are ignitable when they are a mixture of 13% or less by volume with air or (b) have a flammable range with air of at least 12% regardless of the lower limit.

Division 2.2 materials are nonflammable, nonpoisonous, compressed gases including compressed gas, liquefied gas, pressurized cryogenic gas, and compressed gas in solution and are any materials or mixtures which exert in the packaging an absolute pressure of 200 kPa (40 psi) at 20°C (68°F) and do not meet the definition of Division 2.1 and 2.3.

Division 2.3 materials are those which are poisonous and are known to be so toxic to humans as to pose a hazard to health during transportation or, in the absence of adequate data on human toxicity, are presumed to be toxic to humans because when tested on laboratory animals, they have an LC_{50} value not more than 5000 ppm.

Class 3 (Flammable Liquid) Materials A flammable liquid is any liquid having a flash point of not more than 60°C (140°F). Exceptions to this are materials meeting the definition of any Class 2 material; if a mixture has one or more components with a flash point greater than 60.5°C (141°F) or higher that makes up at least 99% of the total volume of the mixture; or a distilled spirit of 140 proof or lower is considered to have a flash point of lower than 23°C (73°F).

A combustible liquid is any liquid that does not meet the definition of any other hazard class which has a flash point above 60.5°C (141°F) and below 93°C (200°F).

Class 4 (Flammable Solid) Materials Class 4 is divided into three divisions as follows:

Division 4.1 includes any flammable solid material of the following three types:

1. Wetted explosives that, when dry, are explosives of Class 1 other than those of compatibility group A, which, when wetted, suppress the explosive properties, and materials specifically authorized by name in 172.101 or by the Associate Administrator for Hazardous Materials
2. Self-reactive materials that are liable to undergo, at normal or elevated temperatures, a strongly exothermal decomposition caused by excessively high transport temperature or by contamination
3. Readily combustible solids that may cause a fire through friction, show a burning rate faster than 2.2 mm (0.087 inches) per second under specified test procedures, or any metal powder that can be ignited and react over the whole length of the sample in 10 minutes or less under specified test procedures

Division 4.2 (spontaneously combustible) material includes:

1. A pyrophoric material, liquid or solid, that even in small quantities and without an external ignition source can ignite within 5 minutes after coming in contact with air under specified test procedures
2. A self-heating material that, when in contact with air and without an energy source, is liable to self-heat and which exhibits spontaneous ignition or which under specified test procedures would be classed as Division 4.2 material

Division 4.3 (dangerous when wet) material is one which, by contact with water, is liable to become spontaneously flammable or which gives off flammable or toxic gas at a rate greater than 1 liter per kilogram of material per hour under specified test procedures

Class 5: Division 5.1 and Division 5.2 Materials *Division 5.1* (oxidizing) material is a material that may, generally by yielding oxygen, cause or enhance the combustion of other materials.

Division 5.2 (organic peroxide) material is an organic compound containing oxygen (O) in the bivalent —O—O— structure and which may be considered a derivative of hydrogen peroxide, where one or more hydrogen atoms have been replaced by organic radicals, with some exceptions.

Class 6 (Poisonous) Materials Class 6 is divided into two divisions as follows:

Division 6.1 materials are (a) those materials, other than gases, which are known to be so toxic to humans as to afford a hazard to health during transportation, or which in the absence of adequate data on human toxicity are presumed to be toxic to humans when they fall within the specified oral, dermal, or inhalation toxicity ranges when tested on laboratory animals, or (b) irritating materials with properties similar to those of tear gas, which causes extreme irritation, especially in confined spaces.

Division 6.2 materials (a) are infectious substances which can be a viable microorganism or its toxin which causes or may cause disease in humans or animals, including those agents listed in 42 CFR 72.3 (infectious substances and etiologic agents); (b) diagnostic specimens—that is, any human or animal material including but not limited to excreta, blood and its components, tissue, and tissue fluids shipped for purposes of diagnosis; or (c) biological product prepared and manufactured in accordance with the provisions of 9 CFR parts 102–104 and 21 CFR parts 312, 600, and 680 and which, in accordance with these provisions, may be shipped in interstate commerce.

Class 7 (Radioactive) Materials Radioactive material is any material having a specific activity greater than 0.002 microcuries per gram.

Class 8 (Corrosive) Materials A corrosive material is a liquid or solid that causes visible destruction or irreversible alteration in human skin tissue at the site of contact, or a liquid that has a severe corrosion rate on steel or aluminum, in accordance with specified criteria.

Class 9 (Miscellaneous) Materials Miscellaneous hazardous material is a material which presents a hazard during transport, but which is not included in any other hazard class. Class 9 includes the following:

1. Any material which has an anesthetic, noxious, or other similar property which could cause extreme annoyance or discomfort to a flight crew member so as to prevent the correct performance of assigned duties
2. Any material that is not included in any other hazard class, but is subject to the requirements of this subchapter because it meets the definition in Section 171.8 of this subchapter for a hazardous substance or hazardous waste

Other Regulated Materials (ORM-D) ORM-D materials such as consumer commodities, which although otherwise subject to the regulations, present a limited hazard during transportation due to their form, quantity, and packaging. It must be material for which exceptions are provided in the Section 172.101 Table.

REVIEW QUESTIONS

1. What is the General Duty Clause, and how can it be used by OSHA to convince employers to correct dangerous situations, when no standards or regulations exist?
2. List two employer responsibilities under the OSH Act.
3. List two employee responsibilities and two employee rights under the OSH Act.
4. What does Title III of SARA mandate?
5. List two materials that can be classified as Class 3, and list two materials that can be classified as Class 9.

REFERENCES

Comprehensive Environmental Response, Compensation and Liability Act (Superfund), PL 96-510, 42 USC Section 9601 et seq., 1980.

DOT 49 CFR171-173: Hazardous Materials Transportation Act, U.S. Government Printing Office, Washington, D.C., 1991.

EPA 40 CFR 300: Comprehensive Environmental Response, Compensation and Liability Act (Superfund), U.S. Government Printing Office, Washington, D.C., 1980.

EPA 40 CFR 240-271: Resource Conservation and Recovery Act (RCRA), U.S. Government Printing Office, Washington, D.C., 1976.

Hazardous Materials Transportation Act (HMTA), PL 93-633, 49 USC 1801 et seq., 1975, 1976, 1981, 1983, 1991.

Levine, S., and W. Martin, eds., *Protecting Personnel at Hazardous Waste Sites*, Butterworth, Stoneham, MA, 1985.

Occupational Safety and Health Act of 1970, PL 91-596, 84 Stat. 1590, Section 6(b)(5).

OSHA, *All About OSHA*, OSHA No. 2056, U.S. Government Printing Office, Washington, D.C., 1991.

OSHA 29 CFR 1910.120: *Hazardous Waste Operations and Emergency Response*, Final Rule, U.S. Government Printing Office, Washington, D.C., 1989.

OSHA 29 CFR 1910.1000: *Air Contaminants (Z-Table)*, U.S. Government Printing Office, Washington, D.C., 1989.

OSHA 29 CFR 1910.134: *Respiratory Protection*, U.S. Government Printing Office, Washington, D.C., 1986.

OSHA 29 CFR 1910.1200: *Hazard Communication*, U.S. Government Printing Office, Washington, D.C., 1983.

OSHA 29 CFR 1910: *Access to Employee Exposure and Medical Records*, U.S. Government Printing Office, Washington, D.C., 1980.

OSHA 29 CFR 1910.38: *Employer Emergency Plans and Fire Prevention Plans*, U.S. Government Printing Office, Washington, D.C., 1980.

OSHA 29 CFR 1910.156: *Fire Brigades*, U.S. Government Printing Office. Washington, D.C., 1980.

Resource Conservation and Recovery Act of 1976, PL 94-580, 42 USC 6901 et seq., 1976.

Superfund Amendments and Reauthorization Act, PL 99-499, October 17, 1986.

U.S. Department of Labor, Title 29 Part 1910, U.S. Government Printing Office, Washington, D.C.

CHAPTER 3

FUNDAMENTALS OF CHEMISTRY

CHAPTER OBJECTIVES

When you have completed this chapter, you will be better able to

- Understand the basic chemistry of hazardous materials
- Balance equations and determine the concentration of solutions
- Understand the differences between acids and bases
- List the heavy metals and their toxicities and sources
- Know the differences between organic and inorganic compounds
- Understand the basic chemistry of organic compounds and their human and environmental impacts

Chemistry, the science of matter and energy, deals with the common everyday things we take for granted: the air we breathe, the water we drink, the soil in which to grow our food, and the vital processes which we define as life.

Chemistry is largely an experimental science. In a broad sense we can view chemistry on three levels. At the first level we observe what actually takes place in an experiment: a rise in temperature, a change in color, the evolution of a gas, and so on. At the second level we record and describe the experiment in scientific language, using shorthand symbols and equations. This shorthand helps to simplify the description and provides a common base by which scientists can communicate with one another. The third level is the process by which we attempt to explain the phenomenon.

The rusting of iron is a common sight. This process is visible to the unaided eye. If we were studying rusting of iron in the laboratory, we could describe the process in chemical shorthand—an equation that tells us how the rust is formed from iron, oxygen, and water, under a given set of conditions. We would address questions like "What actually happens when iron rusts?" and "Why does iron rust while gold does not, under similar conditions?" To answer these and other related questions, we have to know the behavior of the fundamental units of matter, which are atoms and molecules. Because these atoms and molecules are extremely small compared to macroscopic objects such as steel drums, the interpretation of the observed phenomenon takes us into the microscopic world.

The study of chemistry requires that we consider both the macroscopic and the microscopic worlds.

3.1 BASIC CHEMICAL DEFINITIONS

Matter: In the broadest sense, matter is anything that possesses a mass and occupies space.

Mass: Mass is a measure of the quantity of matter in an object.

Weight: Weight refers to the force that gravity exerts on the mass of an object.

Substance: Matter has substance that has a definite or constant composition and distinct properties. Examples include water, air, iron, gold, and oxygen. Substances differ from one another in composition and can be identified by their appearance, smell, taste, and other properties.

Mixture: Mixtures contain two or more substances in variable amounts. Each of the substances retain their identity. For example, air is a mixture of nitrogen, oxygen, carbon dioxide, and several other minor constituents; seawater is a mixture of water, salt, and many other minerals; soil is a complex mixture of many materials. Mixtures are either homogeneous or heterogeneous. When a spoonful of sugar dissolves in water, the composition of the mixture, after sufficient stirring, is the same throughout the solution. This solution is a homogeneous mixture. Oil-and-vinegar salad dressing is an example of a heterogeneous mixture. Oil and vinegar can be thoroughly mixed, yet after standing the two constituents separate by their differences in density.

Physical Property: A property that can be measured and observed without changing the composition or identity of a substance is a physical property. For example, mass, weight, volume, color, size, density, melting and boiling points, and physical state are all physical properties of a substance.

Chemical Property: Those properties of a substance observed when the substance undergoes a chemical reaction are chemical properties. For example, hydrogen reacts with oxygen to form water. Every time we hard-boil an egg for breakfast, we are causing chemical changes. When subjected to a temperature of about 100°C, the yolk and the eggwhite undergo reactions that alter not only their physical appearance but their chemical makeup as well.

Atoms and Elements: Atoms are the smallest units of a substance. All of the millions of substances are composed of approximately 110 elements, each composed of atoms. Each atom of a particular element is chemically identical to every other atom and contains the same number of positively charged units (protons), negatively charged units (electrons), and neutrally charged units (neutrons). The number of protons in the nucleus of each atom of an element is the *atomic number* of the element. The protons and neutrons are located in the nucleus of the atom, while the electrons are located in distinct rings or energy levels around the nucleus. Atomic numbers are integers ranging from 1 to more than 100, each denoting a particular element. In addition to atomic numbers, each element has a name and chemical symbol, such as carbon, C, with an atomic number 6; copper, Cu, atomic number 29; or lead, Pb, atomic number 82. In addition to atomic number, name, and chemical symbol, each element has an *atomic mass* (formerly referred to as *atomic weight*). The atomic mass of an atom is equal to the number of protons (atomic number) plus the number of neutrons present in the nucleus as shown in Figure 3.1. Although atoms of the same element are chemically identical, atoms of most elements consist of two or more isotopes that have different numbers of neutrons in their nuclei. Some isotopes are radioactive, which have unstable nuclei

32 ■ FUNDAMENTALS OF CHEMISTRY

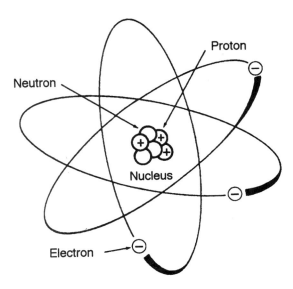

FIGURE 3.1 ■ The basic structure of the atom. Lithium-6 is illustrated.

that give off charged particles and gamma rays in the form of radioactivity. This process of radioactive decay changes atoms of a particular element to atoms of another element.

3.2 THE PERIODIC TABLE

Early scientists in the 19th century realized that a periodic recurrence of chemical properties occurred among the elements. This knowledge led to the development of the periodic table—a tabular arrangement of elements according to increasing atomic number as shown in Figure 3.2.

The horizontal grouping of elements in this table is called a *period*. The vertical grouping of elements in a column of the periodic table is called a *group* or *family*. Resemblance in chemical characteristics occurs for those elements that are members of the same family. There are eight principal families. Four of them deserve special mention:

Group 1A is called the *alkali metals*. Its members are lithium (Li), sodium (Na), potassium (K), rubidium (Rb), cesium (Cs), and francium (Fr). The chemical properties of these elements are strikingly similar. For example, all except lithium explode on contact with water and catch fire easily

1A	2A	3B	4B	5B	6B	7B	8B			1B	2B	3A	4A	5A	6A	7A	8A
1 **H**																	2 **He**
3 **Li**	4 **Be**											5 **B**	6 C	7 **N**	8 **O**	9 **F**	10 **Ne**
11 **Na**	12 **Mg**											13 **Al**	14 **Si**	15 **P**	16 **S**	17 **Cl**	18 **Ar**
19 **K**	20 Ca	21 Sc	22 Ti	23 V	24 Cr	25 Mn	26 Fe	27 Co	28 Ni	29 Cu	30 Zn	31 **Ga**	32 **Ge**	33 **As**	34 **Se**	35 **Br**	36 **Kr**
37 **Rb**	38 Sr	39 Y	40 Zr	41 Nb	42 Mo	43 Tc	44 Ru	45 Rh	46 Pd	47 Ag	48 Cd	49 **In**	50 **Sn**	51 **Sb**	52 **Te**	53 **I**	54 **Xe**
55 Cs	56 Ba	57 La	72 Hf	73 Ta	74 W	75 Re	76 Os	77 Ir	78 Pt	79 Au	80 Hg	81 **Tl**	82 **Pb**	83 **Bi**	84 **Po**	85 **At**	86 **Rn**
87 Fr	88 Ra	89 Ac	104 Unq	105 Unp	106 Unh	107 Uns	108 Uno	109 Une									

58 Ce	59 Pr	60 Nd	61 Pm	62 Sm	63 Eu	64 Gd	65 Tb	66 Dy	67 Ho	68 Er	69 Tm	70 Yb	71 Lu
90 Th	91 Pa	92 U	93 Np	94 Pu	95 Am	96 Cm	97 Bk	98 Cf	99 Es	100 Fm	101 Md	102 No	103 Lr

FIGURE 3.2 ■ Periodic table of the elements. This table shows the positions of the metals, nonmetals, and metalloids. With the exception of hydrogen (H), nonmetals appear at the far right of the table. The elements are arranged according to their atomic mass number.

when exposed to air. Each alkali metal has a characteristic valence or oxidation state of +1 (see Section 3.3 for the explanation of valence).

Group IIA is called the *alkaline earth metals*. Its members are beryllium (Be), magnesium (Mg), calcium (Ca), strontium (Sr), barium (Ba), and radium (Ra). The alkaline earth metals are also chemically reactive, but not nearly as reactive as the alkali metals. They decompose water, but not explosively. They ignite in air, but only after they have been heated or exposed to an ignition source. Each of the alkaline earth metals has a characteristic valence of +2.

Group 7A is known as the *halogens*. Its member are all nonmetals: fluorine (F), chlorine (Cl), bromine (Br), iodine (I), and astatine (At). Each is an exceptionally reactive element. In their simplest compounds, the halogens exhibit a characteristic valence of −1.

Group 8A is known as the *noble*, *rare*, or *inert* gases. Its six members are nonmetals: helium (He), neon (Ne), argon (A), krypton (Kr), xenon (Xe), and radon (Rn). The term *noble* refers to the relatively nonreactive character of the elements.

Hydrogen is considered to be in a family or group by itself. Hydrogen may either lose an electron or gain one. In so doing, it forms the hydrogen (H^+) and hydride (H^-) ions, respectively. Thus, hydrogen has some properties of the alkali metals and some properties of the halogens.

The periodic table is extremely useful for correlating the chemical relationships of the elements. Once a chemical characteristic of one element in a family is known, it can generally be surmised that the other elements in that family possess the same characteristics. Thus, the periodic table represents an important guide for studying the properties of the elements.

3.3 MOLECULES AND CHEMICAL BONDING

Only a few elements exist commonly in nature as individual atoms; most substances exist as molecules, which are aggregates of at least two atoms in a definite arrangement held together by chemical bonds. The bonds may be ionic or covalent, as defined later in this section.

3.3.1 Molecular Formula

A molecular formula shows the exact number of atoms of each element in a molecule. The simplest type of molecule, containing only two atoms, is called a *diatomic molecule*. Elements that exist as diatomic molecules under atmospheric conditions include hydrogen (H_2), nitrogen (N_2), and oxygen (O_2), as well as the group 7A halogens: fluorine (F_2), chlorine (Cl_2), bromine (Br_2), and iodine (I_2). In each case the subscript 2 in the formula indicates the number of atoms in the molecule. Diatomic molecules can contain atoms of two different elements, as in the compounds hydrogen fluoride (HF) and magnesium oxide (MgO). These formulas have no subscripts because when the number of a particular atom is one, the number is not shown as a subscript.

A molecule may contain more than two atoms, either of the same type, as in ozone (O_3), or of different types, as in water (H_2O) and ammonia (NH_3).

The terms "molecule" and "compound" are often used interchangeably, but they do not have the same meaning. A compound is a substance composed of atoms of two or more elements, whereas a molecule is a unit of a substance composed of two or more atoms of the same or different elements. Thus the symbol Cl_2 represents a molecule but not a compound because there is only one type of element present. On the other hand, H_2O_2 represents both a molecule (because there are four atoms present) and a compound (because there are two different elements present).

3.3.2 Ionic Bonding

An ionic bond is formed by the transfer of electrons (from the outer electron shell) from an atom with a low ionization potential, the cation, to an atom with a high electron affinity, the anion (which can accept the electron in its outer electron shell). One atom then takes on a positive charge (the cation) while the other takes on a negative charge (the anion). The ion pair that results is loosely held together by electrostatic attraction. This general rule tells us that the elements most likely to form cations in ionic compounds are the alkali metals and alkaline earth metals, and that the elements most likely to form anions are the halogens and oxygen.

Consider the formation of the ionic compound lithium fluoride from lithium and fluorine. Lithium has a valence of $+1$; that is, it can donate or lose one electron. Fluorine has a valence of -1; that is, it can accept one electron. When the lithium and fluorine atoms come in contact with each other, the valence electron of lithium is transferred to the fluorine atom. Using the Lewis dot symbols (which consist of a symbol for the element and one dot for each valence electron in the atom of the element), we represent the reaction like this:

$$\cdot\text{Li} + :\ddot{\underset{..}{\text{F}}}\cdot \longrightarrow \text{Li} + :\ddot{\underset{..}{\text{F}}}:^- \quad (1)$$

$$\cdot\text{Li} \longrightarrow \text{Li}^+ + e^- \quad (2)$$

$$:\ddot{\underset{..}{\text{F}}}\cdot + e^- \longrightarrow :\ddot{\underset{..}{\text{F}}}:^- \quad (3)$$

Formula 1 depicts the Lewis dot diagrams of the elements of lithium and fluorine before and after ionic bond formation. Formula 2 shows the lithium atom donating an electron to form the lithium ion. Formula 3 shows the fluorine accepting the electron to form the fluoride ion.

Another example is the burning of calcium (Ca) in oxygen (O_2) to form calcium oxide (CaO):

$$2\text{Ca} + O_2 \longrightarrow 2\text{CaO} \quad (4)$$

$$\cdot\text{Ca}\cdot + \cdot\ddot{\underset{..}{\text{O}}}\cdot \longrightarrow \text{Ca}^{2+} \; :\ddot{\underset{..}{\text{O}}}:^{2-} \quad (5)$$

$$\cdot\text{Ca}\cdot \longrightarrow \text{Ca}^{2+} + 2e^- \quad (6)$$

$$\cdot\ddot{\underset{..}{\text{O}}}\cdot + 2e^- \longrightarrow :\ddot{\underset{..}{\text{O}}}:^{2-} \quad (7)$$

Formula 4 depicts the elements of calcium and oxygen before and after formation of calcium oxide (CaO). Formula 5 depicts the Lewis dot diagrams of the elements of calcium and oxygen before and after ionic bond formation. Formula 6 shows the calcium atom donating two electrons to form the calcium ion. Formula 7 shows oxygen accepting the two electrons to form the oxygen ion.

The structures we have shown for lithium, fluorine, calcium, and oxygen are called *Lewis structures*. A Lewis structure is a representation of ionic bonding using Lewis dot symbols in which donated, accepted, or shared electron pairs are shown either as a dot or pairs of dots between two atoms. Only valence electrons are shown in a Lewis structure.

3.3.3 Covalent Bonding

We turn our attention to covalent bonds, an alternate bond that holds atoms together as molecules. Early in this century, Gilbert Lewis discovered the role of electrons in chemical bond formation. In particular, Lewis discovered the concept that a chemical bond involves two atoms in a molecule sharing a pair of electrons. He depicted the formation of a chemical bond in H_2 as

$$H\cdot + H\cdot \longrightarrow H:H \tag{8}$$

This type of electron pairing is an example of a covalent bond, a bond in which two electrons are shared by two atoms. For the sake of simplicity, the shared pair of electrons is often represented by a single line. Thus the covalent bond in the hydrogen molecule can be written as

$$H - H \tag{9}$$

In a covalent bond, each electron in a shared pair is attracted to both of the nuclei involved in the bond. This attraction is responsible for holding the two atoms together. The same type of attraction is responsible for the formation of covalent bonds in molecules other than H_2.

If both atoms forming a covalent bond are the same element (H_2), the electrons will be shared equally. This is known as a *nonpolar covalent bond*. If the atoms are not both the same element, the electrons will not be shared equally, resulting in a *polar covalent bond*. For example, the bond between hydrogen and chlorine in HCl is partially covalent and partially ionic. Thus there is no sharp dividing line between ionic and covalent bonds for most compounds.

When dealing with covalent bonding among atoms other than hydrogen, we need be concerned only with the valence electrons. Consider the fluorine molecule, F_2. The fluorine molecule has seven valence electrons, and the F_2 molecule is represented as

$$:\ddot{F}:\ddot{F}: \quad \text{or} \quad :\ddot{F}-\ddot{F}: \tag{10}$$

Each atom of the fluorine molecule shares the unpaired electron of the other fluorine atom as shown above.

In the water molecule, the electron configuration of the oxygen atom shows that six valence electrons are present, and therefore the oxygen has a valence of -2. The oxygen atom forms two covalent bonds with the two hydrogen atoms in H_2O:

$$H:\ddot{\underset{..}{O}}:H \quad \text{or} \quad H-\underset{..}{\ddot{O}}-O \tag{11}$$

Here we see that the O atom has two lone pairs. The hydrogen atom has no lone pairs because its only electron is used to form the covalent bond.

The structures we have shown for H_2, F_2, and H_2O are called *Lewis structures*. A Lewis structure is a representation of covalent bonding using Lewis dot symbols in which shared electron pairs are shown either as lines or pairs of dots between two atoms, and lone pairs are shown as pairs of dots on individual atoms. Only valence electrons are shown in a Lewis structure.

Two atoms held together by one electron pair are said to be joined by a single bond. In many compounds, however, two atoms share two or more pairs of electrons. Such compounds contain one or more multiple bonds. If two atoms share two pairs of electrons, the covalent bond is called a *double bond*. Double bonds are found in molecules such as carbon dioxide (CO_2)

$$\ddot{\underset{..}{O}}::C::\ddot{\underset{..}{O}} \quad \text{or} \quad \ddot{\underset{..}{O}}=C=\ddot{\underset{..}{O}} \tag{12}$$

and ethylene

$$\begin{array}{cc} H & H \\ \dot{C}::\dot{C} \\ H & H \end{array} \quad \begin{array}{c} H \\ \diagdown\diagup \\ C=C \\ \diagup\diagdown \\ H H \end{array} \tag{13}$$

A triple bond exists when two atoms share three pairs of electrons, as in the nitrogen molecule (N_2):

$$:N\vdots\vdots N: \quad \text{or} \quad :N\equiv N: \tag{14}$$

The acetylene molecule (C_2H_2) also contains a triple bond, in this case between two carbon atoms:

$$H:C\vdots\vdots C:H \quad \text{or} \quad H-C\equiv C-H \tag{15}$$

Note that in ethylene and acetylene all the valence electrons are used in bonding; there are no lone pairs on the carbon atoms.

Ionic and covalent compounds differ markedly in their general physical properties because of the difference in the nature of their bonds. There are two types of attractive forces in covalent compounds. The first type of attractive force operates between molecules and is called an *intermolecular force*. Because intermolecular forces are generally quite weak compared to forces holding atoms together within the molecule, molecules of a covalent compound are not held together tightly. Consequently, covalent compounds are usually gases, liquids, or low-melting solids. On the other hand, the electrostatic forces holding the ions together in an ionic compound are usually very strong, so ionic compounds are always solids with high melting points. Many ionic compounds are soluble in water, and the resulting aqueous solutions conduct electricity, because the compounds are strong electrolytes. Most covalent compounds, on the other hand, are insoluble in water; if they do dissolve, their aqueous solutions do not conduct electricity because the compounds are nonelectrolytes.

3.4 OXIDATION–REDUCTION REACTIONS

We discussed the formation of ionic compounds from metallic and nonmetallic elements in Section 3.3. Let us again consider the reaction between the atoms of lithium and fluorine. For convenience the reaction can be considered as two separate steps, one involving the loss of an electron by the lithium atom and the other the gain of an electron by a fluorine atom:

$$\cdot \text{Li} \longrightarrow \text{Li}^+ + e^-$$

$$:\!\ddot{\text{F}}\!\cdot + e^- \longrightarrow :\!\ddot{\text{F}}\!:^-$$

Each of these steps is called *half-reaction*, which explicitly shows the electrons involved. The sum of these half-reactions gives the overall reaction

$$\cdot \text{Li} + :\!\ddot{\text{F}}\!\cdot + e^- \longrightarrow \text{Li}^+ \; :\!\ddot{\text{F}}\!:^- + e^-$$

or, if we omit the electron that appears on both sides of the equation,

$$\cdot\text{Li} + :\ddot{\text{F}}\cdot \longrightarrow \text{Li}^+ \quad :\ddot{\text{F}}:^-$$

The half-reaction that involves the loss of electrons is called an *oxidation reaction*, the half-reaction that involves the gain in electrons is called a *reduction reaction*. In this example, lithium is oxidized. It is said to act as a *reducing agent* because it *donates an electron* to fluorine and causes fluorine to be reduced. Fluorine is reduced and acts as an *oxidizing agent* because it *accepts an electron* from lithium, causing lithium to be oxidized.

As another example, consider the formation of MgO. The two half-reactions are

$$\cdot\text{Mg}\cdot \longrightarrow \text{Mg}^{2+} + 2e^-$$

$$\cdot\ddot{\text{O}}\cdot + 2e^- \longrightarrow :\ddot{\text{O}}:^{2-}$$

and the overall reaction is

$$\cdot\text{Mg}\cdot + \cdot\ddot{\text{O}}\cdot + 2e^- \longrightarrow \text{Mg}^{2+} + :\ddot{\text{O}}:^{2-} + 2e^-$$

or

$$\cdot\text{Mg}\cdot + \cdot\ddot{\text{O}}\cdot \longrightarrow \text{Mg}^{2+} + :\ddot{\text{O}}:^{2-}$$

In this case, Mg is oxidized and is the reducing agent because it loses two electrons; oxygen is reduced and is the oxidizing agent because it accepts two electrons.

Oxidation and reduction always occur together because in any reaction the total number of electrons lost by the reducing agent must equal the total number of electrons gained by the oxidizing agent. *Redox reaction*, a term that combines "reduction" and "oxidation," is often used for *oxidation–reduction reaction*.

3.4.1 Oxidation State

The oxidation state, oxidation number, or valence of an atom refers to the number of electrons that it gains, loses, or shares with the other atoms.

The loss of two electrons from the magnesium atom, shown above in the formation of magnesium oxide, is an example of oxidation, and the Mg^{2+} ion product is said to be in the +2 oxidation state or have an oxidation number of +2. In the same reaction, the formation of magnesium oxide, the oxygen atom gains two electron and is reduced; it is in the −2 oxidation state or has an oxidation number of −2.

Many hazardous wastes are oxidizers or strong reducers, and the redox reaction is a strong driving force behind many dangerous chemical reactions. For example, the strong oxidizing ability of concentrated sulfuric acid enables it to react violently with any fuel and cause an explosion. The reducing ability of hydrogen and carbon in propane in the presence of an ignition source causes it to burn violently or explode in the presence of the oxygen (oxidizing agent) in air.

The oxidation state (oxidation number or valence) of an element in a compound may have a strong influence on the hazards posed by the compound. For example, chromium compounds with an oxidation state of +3, denoted as chrome (III), are not carcinogenic; however, chromium compounds with an oxidation state of +6 (chrome VI) are known to produce cancer when inhaled.

3.5 ATOMIC AND MOLECULAR MASSES
3.5.1 Atomic Mass

One of the fundamental properties of an atom is its mass. The mass of an atom is related to the number of electrons, protons, and neutrons in the atom. Atoms are extremely small particles—even the smallest speck of dust contains as many as 1×10^{16} atoms. If atoms are so tiny, how do we determine their mass? We cannot weigh a single atom, but there are experimental methods of determining the mass of one atom relative to another.

By international agreement (International System of Units, abbreviated SI, from the French Le Système International d'Unités), an atom of the isotope of carbon (called carbon-12) that has six protons and six neutrons has a mass of exactly 12 atomic mass units (amu). This carbon-12 atom serves as a standard.

Experiments have shown that, on average, a hydrogen atom is only 8.400% as massive as carbon-12. Thus, if we consider the mass of one carbon-12 atom to be exactly 12 amu, then the atomic mass of hydrogen must be $0.008400 \times 12 = 1.008$ amu. Similar calculations show that the atomic mass of oxygen is 16.00 amu and that of iron is 55.85 amu. Note that although we do not know just how much an average iron atom's mass is, we know that it is approximately 56 times as massive as a hydrogen atom.

3.5.2 Molecular Mass

The average mass of all atoms of a compound is the *molecular mass* (formerly called *molecular weight*). The molecular mass of a compound is calculated by multiplying the atomic mass of each element by the relative number of atoms of the element, then adding all the values obtained for each element in the compound. For example, the molecular mass of ammonium hydroxide (NH_4OH) is 14 (N) + 5 (5H) + 16 (O) = 35.

3.5.3 Molar Mass and Avogadro's Number

The mole (mol) is a measure of the quantity of an element or compound. Specifically, a mole of an element will have a mass equal to the element's atomic (or molecular) weight. The SI system has defined the mole as a gram-mole (g mol). Therefore, one mole of carbon-12 equals 12 grams and one mole of calcium equals 40 grams.

One mole of any substance has a number of particles (atoms, molecules, ions, electrons, etc.) equal to 6.022×10^{23}. The number 6.022×10^{23} is called *Avogadro's number* in honor of the Italian scientist Amedeo Avogadro. Therefore, one gram-mole of carbon-12 contains 6.022×10^{23} atoms and has a molecular mass (or weight) of 12 grams. Thus, the atomic mass of sodium (Na) is 22.99 amu, its molar mass is 22.99 grams, and 1 mole contains 6.022×10^{23} atoms.

3.6 CHEMICAL REACTIONS AND BALANCING EQUATIONS

3.6.1 Chemical Reaction

During chemical reactions, bonds between atoms are broken and new bonds are usually formed. The starting substances are known as *reactants*; the ending substances are known as *products*. In a chemical reaction, reactants are either converted to simpler products or synthesized into more complex compounds. There are four common types of reactions:

- *Direct Combination or Synthesis.* This is the simplest type of reaction where two elements or compounds combine directly to form a compound. Consider hydrogen gas burning in air to yield water:

$$2H_2 + O_2 \longrightarrow 2H_2O$$

or the formation of sulfurous acid or acid rain from sulfur dioxide and water in the atmosphere:

$$SO_2 + H_2O \longrightarrow H_2SO_3$$

- *Decomposition.* Bonds within a compound are disrupted by heat or other energy sources to produce simpler compounds or elements.

$$2HgO \longrightarrow 2Hg + O_2$$

$$H_2CO_3 \longrightarrow H_2O + CO_2$$

- *Single Displacement, Substitution, or Replacement.* This type of reaction has one element and one compound as reactants:

$$2Na + 2H_2O \longrightarrow 2NaOH + H_2$$

- *Double Displacement, Substitution, or Replacement.* These are reactions with two compounds as reactants and two compounds as products.

$$AgNO_3 + NaCl \longrightarrow AgCl + NaHO_3$$

$$H_2SO_4 + ZnS \longrightarrow H_2S + ZnSO_4$$

3.6.2 Balancing Chemical Equations

When hydrogen gas is burned in air to form water, the reaction can be represented as

$$H_2 + O_2 \longrightarrow H_2O$$

where the + sign means "reacts with" and the → symbol means "to yield." Thus, this chemical shorthand can be read as follows: "Molecular hydrogen reacts with molecular oxygen to yield water." The reaction is assumed to proceed from the left to the right as the arrow indicates.

This equation is not complete, because there are twice as many oxygen atoms on the left side of the arrow (two) as on the right side (one). To conform to the law of conservation of matter, there must be the same number of each type of atom on both sides of the arrow; that is, we have as many atoms after the reaction as we did before it started. We can balance this equation by placing an appropriate coefficient (2 in this case) in front of the H_2 and H_2O:

$$2H_2 + O_2 \longrightarrow 2H_2O$$

This balanced equation shows that "two diatomic hydrogen molecules can combine or react with one oxygen molecule to form two water molecules. Since the ratio of the number of molecules is equal to the ratio of the number of moles, the equation can also be read as follows: "Two moles of hydrogen react with one mole of oxygen to produce two moles of water." We know the mass of a mole of each of these substances, so we can restate the equation as "4.04 g of H_2 react with 32.00 g of O_2 to give 36.04 g of H_2O."

Now let us consider the combustion of ethane (C_2H_6) in air, which yields carbon dioxide (CO_2) and water (H_2O). We write

$$C_2H_6 + O_2 \longrightarrow CO_2 + H_2O$$

We can see that the equation is not balanced. By deductive trial and error, we adjust the coefficients to balance the equation:

$$2C_2H_6 + 7O_2 \longrightarrow 4CO_2 + 6H_2O$$

3.7 SOLUTIONS

In a broad sense, solutions are homogeneous mixtures consisting of two or more substances in the same phase. By this definition, mixtures of gases are solutions because the component gases are homogeneously distributed. Similarly, a solution of solids can be formed in which a substance can be homogeneously distributed throughout the solid. Examples of solid solutions are sterling silver (silver with added copper) and brass (copper with added zinc). We usually think of a solution as a liquid because water-based solutions are the most familiar.

A solution is considered to be made up of two types of components. The component present in the larger amount is called the *solvent*, and any substance or substances present in the smaller amount is the *solute*. More than one solute can be present in the solution at the same time.

When a solid is added to a liquid, the solid is known as the *solute* and the liquid is known as the *solvent*. If the dispersion of the solute throughout the solvent is at the molecular level, the mixture is known as a *solution*. If the solute particles are larger than molecules, the mixture is known as a *suspension*.

There may be no readily visible evidence that a solution exists. For example, a deadly solution of sodium cyanide in water looks like pure water. The solution may have a strong color, as is the case for intensely purple solutions of potassium permanganate, $KMnO_4$. It may have a strong odor, such as that of ammonium hydroxide, NH_4OH, dissolved in water. Solutions may consist of solids, liquids, or gases dissolved in a solvent.

3.7.1 Solution Concentration

Solutions are used in many chemical processes, and many chemical reactions occur in solution. Therefore, it is of importance to be able to quantify the amount of solute relative to solvent. The concentration of a solution is the amount of solute dissolved in a given quantity of solvent. There are numerous units of concentration used by industry. Several of these will be defined below.

Percent by Weight Concentrations are expressed in numerous ways. Very high concentrations are often given as a percent by weight. For example, commercial concentrated hydrochloric acid is 36% HCl, meaning that 36% of the weight comes from the dissolved HCl (solute) and 64% from the water (solvent). Percent by weight can also be defined as

Percent by weight of solute

$$= \frac{\text{Weight of solute}}{\text{Weight of solute} + \text{Weight of solvent}} \times 100\%$$

$$= \frac{\text{Weight of solute}}{\text{Weight of solution}} \times 100\%$$

For example, a sample of 0.892 g of potassium chloride (KCl) is dissolved in 54.6 g of water. What is the percent by weight of KCl in solution?

Percent by weight of KCl

$$= \frac{\text{Weight of solute}}{\text{Weight of solute} + \text{Weight of solvent}} \times 100\%$$

$$= \frac{0.892 \text{ g}}{0.892 \text{ g} + 54.6 \text{ g}} \times 100\%$$

$$= 1.61\% \text{ weight (\%W)}$$

Molarity One of the most common units of concentration in chemistry is molarity, symbolized by M and also called molar concentration. Molarity is the number of moles of solute in 1 liter of solution at 25°C. Molarity is defined by the equation

$$M = \text{Molarity} = \frac{\text{Moles of solute}}{\text{Liters of solution}}$$

Thus a 3.40 molar solution of potassium nitrate (KNO_3) solution, expressed as 3.40 M KNO_3, contains 3.40 moles of solute (KNO_3) in 1 liter of solution; a 1.66 molar glucose ($C_6H_{12}O_6$) solution, expressed as 1.66 M $C_6H_{12}O_6$ contains 1.66 M $C_6H_{12}O_6$ (the solute) in 1 liter of solution. Of course, we do not always need to work with solutions of 1 liter. Thus, a 500-ml solution containing 1.70 mol of KNO_3 still has the same concentration of 3.40 M:

$$M = \text{Molarity} = \frac{\text{Moles of solute}}{\text{Liters of solution}}$$

$$= \frac{1.70 \text{ mol}}{0.500 \text{ liter}}$$

$$= 3.4 \text{ mol/liter} = 3.40 \text{ M}$$

Milligrams per Liter, Parts per Million (ppm), and Parts per Billion (ppb) Concentrations of dilute solutions, such as spent plating bath rinse solutions, can be expressed as weight of solute per unit volume of solution. Common units used are milligrams per liter (mg/liter) or micrograms per liter (µg/liter). Since a liter of water weighs essentially 1000 grams, a concentration of 1 mg/liter is approximately equal to 1 part per million (ppm) and a concentration of 1 µg/liter is approximately equal to 1 part per billion (ppb).

3.7.2 Electrolytes

Electrolytes Earlier we noted that water is generally a good solvent for ionic compounds. One method of classifying solutions is based on whether a solution made from that solute can conduct electricity. Solutions can conduct electricity when ions are present to carry the electric current through the solution. Positively charged ions (cations) move to the negatively charged pole or anode of the electrical field and negatively charged ions (anions) move toward the positive pole or cathode. If the solution is found to conduct electricity, the solute is called an *electrolyte*. If it does not, the solute is a *nonelectrolyte*.

Electrolytes form solutions that conduct electricity because they dissociate; that is, they break into parts. When an electrolyte dissolves in the solvent, the electrically neutral solute breaks apart to form at least one cation and at least one anion. Solutes that dissolve but do not dissociate are nonelectrolytes—for example, sucrose (table sugar) in water.

Not all electrolytes dissociate to the same extent. Some, called *strong electrolytes*, dissociate completely when they are dissolved. Sodium hydroxide, hydrogen chloride, and calcium chloride are all strong electrolytes in water. When dissolved in water, they dissociate as follows:

$$NaOH \xrightarrow{H_2O} Na^+ + OH^-$$

$$HCl \xrightarrow{H_2O} H^+ + Cl^-$$

$$CaCl_2 \xrightarrow{H_2O} Ca^{2+} + 2Cl^-$$

Strong electrolytes generally fit into three categories: strong acids, strong bases, and most of the common salts. In the three examples cited above, NaOH is a strong base, HCl is a strong acid, and $CaCl_2$ is a salt.

Weak electrolytes, on the other hand, only partially dissociate on dissolving, giving rise to solutions that are poorer conductors of electricity than those of strong electrolytes. A solution of a weak electrolyte does contain dissociated ions, but most of the solute remains undissociated. Consider formic acid, $HCHO_2$: In a solution made by diluting 1 mol of $HCHO_2$ to 1 liter with water, $HCHO_2$ is less than 2% dissociated. Almost all of the $HCHO_2$ remains in the molecular form. Thus the reaction is as follows:

$$HCHO_2 \xrightarrow{H_2O} HCHO_2 + H^+ + CHO_2$$

Nonelectrolytes are solutes that do not dissociate in solution. Examples of nonelectrolyte solutions are ethanol (C_2H_5OH) dissolved in water, table sugar ($C_{12}H_{22}O_{11}$) dissolved in water, and oil dissolved in gasoline.

3.8 ACIDS, BASES, AND NEUTRALIZATION

3.8.1 Ionization of Water, Acids, and Bases

Water is a molecular substance, and therefore we would expect liquid water to be nonconducting. But in fact pure water does conduct small amounts of electricity. Water molecules dissociate very weakly to form a very small number of H^+ and OH^- ions:

$$H_2O \leftrightarrows H^+ + OH^-$$

At 25°C, the concentration of H^+ and OH^- ions is only 1.0×10^{-7} M, whereas the concentration of water molecules is 55.6 M. In other words, only about two out of every billion water molecules are dissociated.

The molar concentrations of hydrogen ion, $[H^+]$, range over several orders of magnitude and are expressed by pH defined as

$$pH = -\log [H^+]$$

In pure water the value of $[H^+]$ is defined exactly to be 1×10^{-7} mol/liter, the pH is 7.00, and the solution is neutral (neither acidic nor basic). Acidic solutions have pH values less than 7, and basic solutions have pH values greater than 7. The lower the pH, the more acidic the solution; the higher the pH, the more basic the solution (see Figure 3.3).

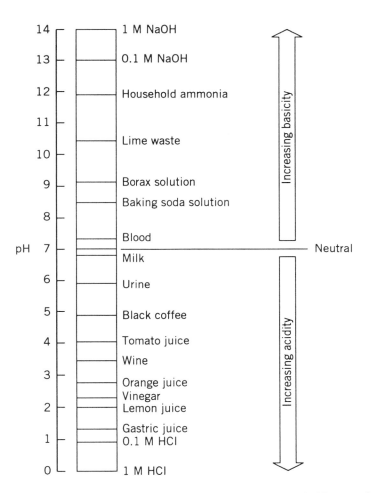

FIGURE 3.3 ■ pH values for several common solutions. The pH scale is 0–14 as shown.

Strong acids and strong bases are corrosive substances that exhibit extremes of pH. They are destructive to many materials, including steel and flesh.

3.8.2 Acid and Base Definitions

Pure water, as we have seen, contains small amounts of H^+ and OH^- ions. We also know that some solutes, when added to water, can affect H^+

and OH^- concentration of the solution. These substances are categorized as acids or bases, depending on how they alter the concentrations of H^+ and OH^-.

Acids In general, acids are molecular substances that produce hydronium ions by reaction with water. For example, hydrogen chloride is an acid, because when it is dissolved in water it reacts with the solvent to produce H_3O^+

$$HCl + H_2O \longrightarrow H_3O^+ + Cl^-$$

If we use H^+ as an abbreviation for the hydronium ion and leave out the molecule of water that carries the H^+, we can also write the reaction as

$$HCl \longrightarrow H^+ + Cl^-$$

Aqueous solutions of hydrogen chloride are called *hydrochloric acid* (HCl). As you have learned, HCl is a strong acid and therefore a strong electrolyte. Its reaction is complete; all of the HCl dissociates. In other words, 1 mol of HCl releases 1 mol of H^+.

Many acids are also weak electrolytes. Acetic acid, $HC_2H_3O_2$, is an example. Acetic acid reacts with water to produce

$$HC_2H_3O_2 \rightleftharpoons H^+ + OC_2H_3O_2^-$$

This is an equilibrium, as indicated by the two arrows. The size and direction of the arrows indicate the major direction of the reaction and in this example, in the solution of acetic acid only a small fraction of the solute is dissociated into ions. This means that the concentration of H^+ ions in solution is low; therefore, the pH is low. As a result, acetic acid and other acids that are weak electrolytes are called *weak acids*.

Bases A base is any substance that, when dissolved in water, increases the concentration of hydroxide ion, OH^-.

$$B + H_2O \longrightarrow BH^+ + OH^-$$

Sodium hydroxide, NaOH, is a strong base. It dissolves and dissociates completely in water:

$$\text{NaOH} + \text{H}_2\text{O} \longrightarrow \text{Na}^+ + \text{OH}^-$$

Ammonia, NH_3, is a weak base. It reacts with water in an equilibrium because only a small fraction of the NH_3 in solution is present as NH_4^+ and OH^- at any given instant:

$$\text{NH}_3 + \text{H}_2\text{O} \rightleftarrows \text{NH}_4^+ + \text{OH}^-$$

Strong and Weak Acids and Bases Strong acids dissociate (ionize) completely in water. Weak acids only partially dissociated (ionize) as a result of equilibrium reactions with water.

Strong bases are present in aqueous solution entirely as ions (completely dissociated), one of which is OH^-. Weak bases only partially ionized as the result of equilibrium reactions with water.

Since both acids and bases are corrosive materials, they can cause skin and eye injury to site workers by direct contact and must be handled carefully. In addition, these materials must be containerized properly to avoid corrosion of the container, and therefore they have to be identified in the field. Figure 3.4 shows a rapid field method to measure the pH of samples with pH paper. The change in color of the paper is compared to the color scale provided with the pH paper.

3.8.3 Neutralization

The most important reaction that acids and bases undergo is their reaction with each other, a reaction called *neutralization*. In aqueous solutions, when a strong acid reacts with a strong base, the products are water, a solution of a salt, and heat:

$$\text{HCl} + \text{NaOH} \longrightarrow \text{H}_2\text{O} + \text{NaCl} + \Delta$$

The source of the H^+ ions is HCl. The source of the OH^- ions is NaOH. The salt, NaCl, and the H_2O are the products of the neutralization reaction. The salt is composed of Na^+ ions and Cl^- ions held together by ionic bonds. Delta (Δ) symbolizes the heat liberated by the reaction.

The solution produced by the reaction of equal amounts of a strong acid and a strong base is a *neutral solution*.

52 ■ FUNDAMENTALS OF CHEMISTRY

FIGURE 3.4 ■ **Testing of field samples for pH with pH paper.**

Neutralization reactions are essential to many industries. Their importance is reflected in the huge amounts of acids and bases produced in the United States each year. In 1990, in the United States alone, more than 90 billion pounds of sulfuric acid and 25 billion pounds of sodium hydroxide were produced. Of the top 15 chemicals produced in the United States in 1990, more than seven are common acids and bases.

The production of sulfuric acid far exceeds that of any other synthetic chemical substance. It is used in the manufacture of phosphate fertilizers, paints and pigments, ammonium sulfate, and rayon as well as in the pickling of iron and steel and in the refining of petroleum. Phosphoric acid and nitric acid are used extensively in the fertilizer industry as well as in the production of other chemicals. One of the major uses of lime is as a flux for steel. Sodium hydroxide, also called *caustic soda*, is used primarily as a base in the manufacture of many chemicals and in the pulp and paper industry. Sodium carbonate, or soda ash, is used in the manufacture of glass and in the pulp and paper industry.

These chemicals and the wastes derived from their uses can be found on many potential remediation sites. Commonly found as soil and groundwater contaminants, they are the result of continuous, proper and improper, legal and illegal disposal of the chemicals at these sites.

3.9 TRACE INORGANIC CONTAMINANTS AND THE HEAVY METALS

The principal source of almost all trace inorganic contaminants is industrial wastes from manufacturing or metal finishing operations. Many of these industrial waste streams have been and still are being discharged into municipal wastewater treatment facilities. In the past, many publicly owned treatment works (POTWs) did not reduce the heavy metals to acceptable concentrations before discharge into surface or groundwaters. Today, treatment at the source is required prior to discharge to the municipal system.

The heavy metals are considered hazardous because of their toxicity. The following elements are included in this group.

> *Arsenic.* Arsenic is a toxic metalloid element. It is a confirmed human carcinogen and can cause both acute and chronic poisoning. Arsenic is a constituent of the earth's crust at an average concentration of 2–5 ppm. The combustion of fossil fuels, particularly coal, introduces large quanitities of arsenic into the environment. Arsenic was used formerly in the formulation of pesticides that were highly toxic. Another main source of arsenic is mine tailings. Arsenic produced as a by-product of copper, gold, and lead refining greatly exceeds the commercial demand for arsenic, and it accumulates as a waste material.
>
> *Barium.* Barium salts are used in the manufacture of paints, linoleum, paper, and drilling muds. Animal tests have shown that prolonged exposure to barium produces muscular and cardiovascular disorders and kidney damage.
>
> *Beryllium.* Beryllium is used in metallurgy to make special alloys, and the oxide is used to make ceramics with electrical applications. The inhalation of beryllium dusts has been found to cause cancer in animals, and for that reason it is considered a possible human carcinogen.
>
> *Cadmium.* Cadmium is used extensively in the manufacture of batteries, paints, and plastics. In addition, it is used in the electroplating

of iron products to prevent corrosion. Most of the cadmium contamination has reached the environment through its use in the metal finishing industry. At low levels, cadmium can cause high blood pressure, sterility among males, and kidney damage; at very high levels, it can cause an illness called "Itai-Itai" disease, characterized by brittle bones and intense pain.

Chromium. Because of its corrosion and wear-resistant properties, chromium is used in the formulation of stainless steels and surface plating of many metals. Chromium sulfate is used in some leather tanning processes and in wood preservation. Chromium exists in several oxidation states or valences in the water environment. The trivalent (+3) and hexavalent (+6) oxidation states are predominant. No evidence at this time indicates that the trivalent form is toxic or a health risk. However, the hexavalent form is a confirmed human carcinogen. Chromate poisoning can cause skin disorders and liver damage.

Lead. Lead is highly toxic and is considered a suspected human carcinogen by the American Conference of Government Hygienists. Lead from leaded gasoline is the major source of lead pollution of the atmosphere and soil. Other sources include the lead used in paints, as well as the manufacture and disposal of storage batteries.

Mercury. Mercury enters the environment as a by-product in the manufacture of numerous products including batteries, thermometers, scientific equipment, fungicides, fumigants, pharmaceutical products, and amalgams. Among the more severe toxicological effects of mercury are neurological damage, including irritability, paralysis, blindness, or insanity ("madhatter's syndrome"); chromosome breakage; and birth defects. Milder symptoms of mercury poisoning include depression and irritability.

Nickel. Nickel is used in electroplating operations as a constituent of corrosion-resistant coatings. The spent plating baths and rinse waters constitute a major avenue by which salts of these metals gain access to the environment. Nickel sulfide dust is a human carcinogen when inhaled. Otherwise, nickel appears to be of low toxicity.

Selenium. Selenium at low levels is essential in the diet of humans and many animals, but at higher concentrations it can be quite toxic.

Selenium is used in chromium plating and in the production of iron, copper, and lead metals and their alloys. Some photoelectric cells and rectifiers contain selenium. Inorganic selenium compounds are used in photocopying processes and in the manufacture of pigments, ceramics, and glass.

Thallium. The industrial applications of thallium and its compounds have been limited because of their high toxicities. Thallium sulfate is used as a pesticide to control animal pests and crawling insects. Besides its use as a rodenticide, thallium is used in semiconductor research and for making alloys with mercury for use in switches operating at subzero temperatures.

All of the heavy metals are constituents of the listed hazardous wastes as described in Section 3001 of RCRA.

3.10 ORGANIC CHEMISTRY

Organic chemistry originally meant the study of substances derived from living systems in nature, whereas inorganic chemistry meant the study of mineral constituents of the earth.

This definition of organic chemistry changed in 1824, when Friedrich Wöhler synthesized urea from inorganic compounds. Since that time many organic compounds have been prepared synthetically in the laboratory and in industry.

All organic compounds contain carbon in combination with one or more elements. The hydrocarbons contain only hydrogen and carbon. A great many compounds contain carbon, hydrogen, and oxygen, and they are considered to be major elements. Minor elements in naturally occurring compounds are nitrogen, phosphorus, and sulfur, and sometimes halogens and metals. Compounds produced synthetically may contain, in addition, a wide variety of other elements.

3.10.1 Properties of Organic Compounds

In general, the properties of organic compounds differ from those of inorganic compounds in several respects:

- Organic compounds are usually combustible.
- Organic compounds usually have lower melting and boiling points.
- Organic compounds are usually less soluble in water.
- Several organic compounds may exist for a given formula. This property is known as *isomerism*.
- Reactions of organic compounds are usually molecular rather than ionic. As a result, they are often quite slow.
- The molecular weights of organic compounds may be very high, often well above 1000 amu.
- Most natural organic compounds can be broken down by bacteria.

Sources
- *Nature*: Fibers, animal and vegetable oils, alkaloids, cellulose, starch, sugars, and so on.
- *Synthesis*: A large number and variety of compounds and materials are produced by manufacturing processes.
- *Fermentation*: Alcohols, acetone, glycerol, antibiotics, acids, and so on, are derived by the action of microorganisms upon organic matter.

The wastes produced in the processing of organic materials and from the synthetic organic and fermentation industries constitute a major part of the industrial and hazardous waste contamination that is the subject of remediation efforts underway at many Superfund and other cleanup sites.

3.10.2 The Carbon Atom

An enormous number of different organic compounds exist as a consequence of the ability of carbon atoms to bond to each other in a limitless variety of straight chains, branched chains, and ring structures.

In inorganic chemistry, a molecular formula is specific for one compound. In organic chemistry, most molecular formulas do not represent any particular compound. For example, the molecular formula $C_3H_6O_3$ represents at least four separate compounds and therefore is of little value in imparting information other than that the compound contains carbon, hydrogen, and oxygen. Four compounds having the structural or molecular formula $C_3H_6O_3$ are

$$\begin{array}{c} \text{H} \\ | \\ \text{H} \quad \text{O} \quad \text{O} \\ | \quad | \quad \| \\ \text{H}-\text{C}-\text{C}-\text{C}-\text{O}-\text{H} \\ | \quad | \\ \text{H} \quad \text{H} \end{array} \qquad \begin{array}{c} \text{H} \quad \text{H} \quad \text{O} \\ | \quad | \quad \| \\ \text{H}-\text{O}-\text{C}-\text{C}-\text{C}-\text{O}-\text{H} \\ | \quad | \\ \text{H} \quad \text{H} \end{array}$$

$$\begin{array}{c} \text{H} \quad \text{O} \quad \text{H} \\ | \quad \| \quad | \\ \text{H}-\text{O}-\text{C}-\text{C}-\text{O}-\text{C}-\text{H} \\ | \quad \quad | \\ \text{H} \quad \quad \text{H} \end{array} \qquad \begin{array}{c} \text{H} \quad \text{H} \quad \text{O} \\ | \quad | \quad \| \\ \text{H}-\text{C}-\text{O}-\text{C}-\text{C}-\text{OH} \\ | \quad | \\ \text{H} \quad \text{H} \end{array}$$

Compounds having the same molecular formula are known as *isomers*. In many cases, structural formulas may be simplified or condensed. Thus, the formula for propanol

$$\begin{array}{c} \text{H} \quad \text{H} \quad \text{H} \\ | \quad | \quad | \\ \text{H}-\text{C}-\text{C}-\text{C}-\text{O}-\text{H} \\ | \quad | \quad | \\ \text{H} \quad \text{H} \quad \text{H} \end{array}$$

may be written as

$$CH_3 - CH_2 - CH_2OH \quad \text{or} \quad CH_3CH_2CH_2OH$$

This saves both space and effort.

There are two major types of organic compounds, *aliphatic* and *aromatic*. The aliphatic compounds are those in which the characteristic groups are linked to a straight or branched carbon chain. The aromatic compounds have these groups linked to a particular type of six-member carbon ring that contains three double bonds, the benzene ring. Such rings have peculiar stability and chemical character. They are present in a wide variety of important compounds.

3.10.3 Aliphatic Hydrocarbons

Hydrocarbons are compounds of carbon and hydrogen. There are two main types, saturated and unsaturated. Saturated hydrocarbons are those in which the adjacent carbon atoms are joined by a single covalent bond; all other bonds are to hydrogen atoms. Propane is an example of a saturated aliphatic hydrocarbon:

$$\begin{array}{c} \text{H} \quad \text{H} \quad \text{H} \\ | \quad | \quad | \\ \text{H}-\text{C}-\text{C}-\text{C}-\text{H} \\ | \quad | \quad | \\ \text{H} \quad \text{H} \quad \text{H} \end{array} \quad \text{or} \quad CH_3CH_2CH_3$$

Unsaturated hydrocarbons have at least two carbons that are joined by more than one covalent bond; all the remaining bonds are to hydrogen atoms. The following examples are of unsaturated aliphatic compounds with a double bond (propylene) and a triple bond (acetylene), respectively:

$$\begin{array}{c} \text{H} \quad \text{H} \quad \text{H} \\ | \quad | \quad | \\ \text{H}-\text{C}-\text{C}=\text{C}-\text{H} \\ | \quad | \\ \text{H} \quad \text{H} \end{array} \quad \quad HC \equiv CH$$

Alkanes Saturated aliphatic hydrocarbons form a whole series of compounds, starting with one carbon and increasing one carbon, stepwise. These compounds are known as the *paraffin* or *methane* series and as *alkanes*. The principle source is petroleum. Gasoline is a mixture of several of them; diesel fuel is another such mixture.

Methane (CH_4) is the simplest hydrocarbon. It is a gas of importance to remediation workers because it is a major end product of anaerobic digestion of organic materials in landfills. Methane is also a component of marsh gas and natural gas. In a mixture with air containing 5–15% methane, it is highly explosive.

Ethane (CH_3—CH_3) is the second member of this series.

Propane (CH_3—CH_2—CH_3) is the third member of the series.

Butane (C_4H_{10}), the fourth member of the series, occurs in two isomeric forms:

$$H-\underset{\underset{H}{|}}{\overset{\overset{H}{|}}{C}}-\underset{\underset{H}{|}}{\overset{\overset{H}{|}}{C}}-\underset{\underset{H}{|}}{\overset{\overset{H}{|}}{C}}-\underset{\underset{H}{|}}{\overset{\overset{H}{|}}{C}}-H \qquad H-\underset{\underset{H}{|}}{\overset{\overset{H}{|}}{C}}-\underset{\underset{\underset{\underset{H}{|}}{\overset{\overset{H}{|}}{C}-H}}{|}}{\overset{\overset{H}{|}}{C}}-\underset{\underset{H}{|}}{\overset{\overset{H}{|}}{C}}-H$$

n-Butane   Isobutane

Pentane, the fifth member of the series, exists in three isomeric forms.

The methane series of hydrocarbons is characterized by names ending in *-ane*. These straight-chain or branched-chain compounds are alkanes.

Chemical Reactions of Alkanes   At elevated temperatures, strong bases, acids, or oxidizing agents such as sulfuric acid react often explosively with the alkanes to yield carbon dioxide and water.

Alkenes   The members of this group of chemicals are also called *olefins* and belong to the ethylene series. Members of this compounds have names ending in *-ene*. The alkene compounds, particularly ethylene, propylene, and butylene, are formed in great quantities during the "cracking" or pyrolysis of petroleum.

Alkenes are hydrocarbons that have at least one carbon–carbon double bond. Because alkenes can be bonded to additional atoms, they are called *unsaturated hydrocarbons*. The general formula for noncyclic alkanes is C_nH_{2n+2}. The simplest alkene is ethylene or ethene, which has the structure

$$\underset{H}{\overset{H}{\diagdown}}C=C\underset{\diagdown H}{\overset{\diagup H}{}}$$

Alkenes can contain both single and double carbon–carbon bonds. For example, the alkene called *propene*

$$\text{H}_3\text{C}\diagdown\text{C}=\text{C}\diagup\text{H} \quad \text{(with H above left C and H below right C)}$$

has a methyl group in place of one of the hydrogens of ethylene. Noncyclic alkenes that have only one double bond have the general formula C_nH_{2n}.

Alkynes The alkynes are another class of unsaturated hydrocarbons that contain at least one carbon–carbon triple bond. When a noncyclic alkyne has only one triple bond, it has the general formula C_nH_{2n-2}. The simplest alkyne has $n = 2$, so it has the formula C_2H_2. The simplest alkyne is

$$H-C\equiv C-H$$

Cyclic Aliphatic Hydrocarbons These compounds, also called *cyclo-compounds*, have their carbon atoms arranged in ringlike structures. All three kinds of aliphatic hydrocarbons—alkanes, alkenes, and alkynes—can form cyclo-structures. The simplest cycloalkane contains three carbon atoms. Its name is therefore based on the compound propane, and is called cyclopropane. It has the formula C_3H_6 and the structure

As the formula for cyclopropane indicates, a cycloalkane has the same general formula as an alkene containing one double bond, C_nH_{2n}. An example of a cycloalkene is cyclopentene:

3.10.4 Aromatic Hydrocarbons

Aromatics represent a special class of unsaturated ring compounds. Two series of homologous aromatic hydrocarbon compounds are known: the benzene and polyring series.

The simplest member of the benzene series of aromatic hydrocarbons is benzene, C_6H_6. Benzene is a highly unsaturated ring compound that has the following structures:

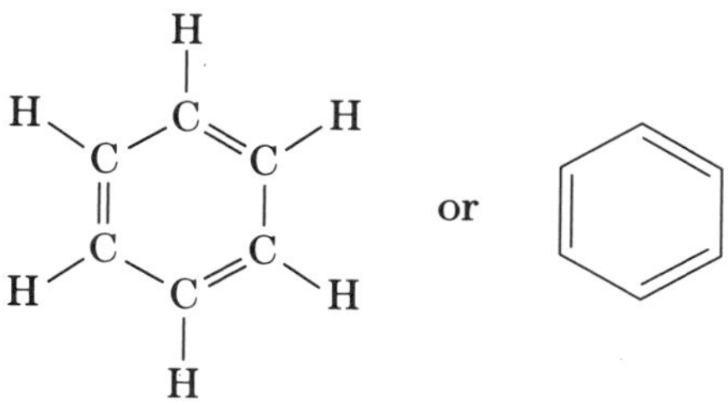

Members of the benzene series are found along with benzene in coal tar and in many crude petroleums. Table 3.1 lists the benzene series hydrocarbons of commercial importance. Toluene, or methyl benzene, is the simplest alkyl derivative of benzene. Xylene is a dimethyl derivative of benzene. It exists in three isomeric forms: ortho-xylene, meta-xylene, and para-xylene. Ethyl benzene is an ethyl derivative of benzene and has the same general formula as the xylenes, C_8H_{10}.

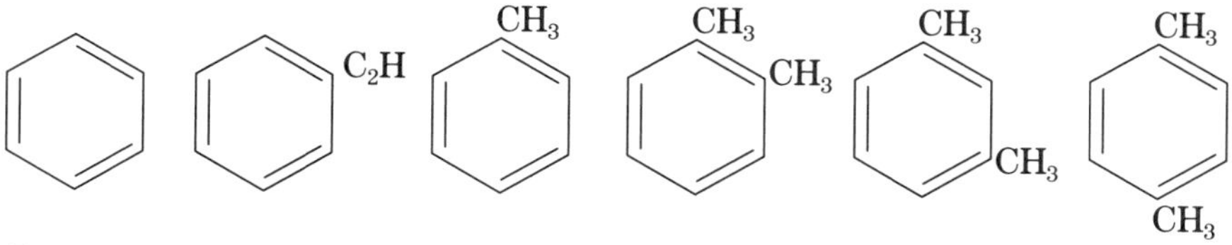

Benzene　Ethylbenzene　Toluene　o-Xylene　m-Xylene　p-Xylene

These compounds are commonly referred to as the *BTEX group*.

The benzene-series hydrocarbons, which are used extensively as solvents and in chemical synthesis, are common constituents of petroleum products (gasoline). Although they are relatively insoluble in water, wastewaters and leachates containing 10 to as high as 1000 mg/liter for BTEX have been observed. These compounds are frequently detected in groundwater. A major source is leaking underground gasoline storage tanks.

TABLE 3.1 ■ Benzene-Series Hydrocarbons

Name	Formula	Flash Point
Benzene	C_6H_6	12°F
Toluene	$C_6H_5 \cdot CH_3$	40°F
Ethylbenzene	$C_6H_5 \cdot C_2H_5$	55°F
o-Xylene	$C_6H_4 \cdot (CH_3)_2$	63°F
m-Xylene		84°F
p-Xylene		81°F

The benzene-series compounds have been implicated in several human health effects, most notably cancer. Benzene is a known cause of leukemia. The current federal drinking water maximum contamination levels (MCLs) are 5 µg/liter for benzene, 700 µg/liter for ethylbenzene, 1 mg/liter for toluene, and 10 mg/liter for the xylenes.

Styrene, another benzene derivative, is also an environmentally significant aromatic compound. It is used as a monomer in the manufacture of a wide variety of polystyrene products (e.g., plastics and synthetic rubber). A drinking water MCL of 100 µg/liter has been set for styrene.

Styrene

Polyring Hydrocarbons A wide variety of polycyclic aromatic hydrocarbons (PAHs) are known. A few of them will illustrate their structure, function, and problems in relation to the environment.

Naphthalene ($C_{10}H_8$) Naphthalene, a white crystalline compound derived from coal tar, was used in mothballs. It is the simplest member of a large number of PAHs having two or more fused rings. The chemical structures for several PAHs are as follows:

Naphthalene   Anthracene   Phenanthrene

Anthracene ($C_{14}H_{10}$) and Phenanthrene ($C_{14}H_{10}$). Anthracene and phenanthrene are isomers. Their formulas illustrate the possible ways in which PAH may occur. Anthracene and phenanthrene are widely used in the manufacture of dyestuffs. The phenanthrene nucleus is found in many important alkaloids, such as morphine, vitamin D, sex hormones, and other compounds of great biological significance.

PAHs are associated with many combustion and pyrolysis processes; therefore, these compounds are encountered abundantly in the atmosphere, soil, and elsewhere in the environment from sources that include engine exhaust, wood stove smoke, cigarette smoke, and char-broiled food. Coal tars and petroleum residues such as road and roofing asphalt have high levels of PAHs. Some PAH compounds, including benzo(a)pyrene, are of medical concern because they are precursors to cancer-causing metabolites.

PAHs are believed to be the cause of the first recognized chemically related cancer which was found in chimney sweeps in the late 18th century. They are common residual contaminants at sites where coal was used to manufacture consumer gas in the early 20th century and where creosote, which is manufactured from coal, was used to preserve wood. The larger compounds (five or more aromatic rings) tend not only to be carcinogenic but also the more difficult for bacteria to degrade.

Chlorinated Aromatic Hydrocarbons (Aryl Halides) There are many industrially important chlorinated aromatic compounds. Their widespread use and environmental persistence have become significant environmental problems, much in the same manner as the chlorinated aliphatic compounds. Two of the most important classes are the chlorinated benzenes and the polychlorinated biphenyls.

Chlorinated Benzenes These chemicals are benzenes with one or more of the hydrogen atoms replaced with chlorine atoms. They are widely used in the chemical industry as solvents or as precursors for pesticide production. Like the chlorinated aliphatics, they have been found in abundant quantities at abandoned waste sites and in many wastewaters and leachates. They are fairly volatile, and they are slightly to moderately soluble in aqueous solutions. The structural diagrams describes several of the chlorinated benzenes and their isomers.

Monochlorobenzene   1,2-Dichlorobenzene   1,4-Dichlorobenzene

1,2,4-Trichlorobenzene   Hexachlorobenzene

Polychlorinated Biphenyls (PCBs) When the hydrogen atoms in biphenyl are replaced with chlorine atoms, polychlorinated biphenyls are produced. They have many desirable physical and chemical properties that have led to their widespread use in many consumer products. They are excellent flame retardants, and as oils or lubricants they have been used in many types of electric motors: refrigerators, air conditioners, clothes washers, dryers, and furnace blowers. They have been used in electrical transformers and capacitors such as those seen on utility poles that feed electricity into homes and factories. They have also been widely used as plasticizers, solvents, and hydraulic fluids. PCBs are highly stable and hydrophobic. They tend to bioconcentrate and therefore to persist in the environment. Significant concentrations of PCBs have been found in the higher levels of the food chain (e.g., fish and birds).

For more than 50 years, large quantities of these stable PCBs have found their way into the environment. They are now considered extremely hazardous, and because of the possible human health effects and their environmental impact, their manufacture in the United States was banned in 1977. The drinking water MCL for total PCB is 0.5 μg/liter.

The disposal of PCBs is strictly controlled and continues to pose environmental problems. Incineration has been used successfully. However, it is a costly method of disposal and has caused many public relations problems. Some PCBs have been shown to be biodegraded by microorganisms: the more highly chlorinated ones by anaerobic microbes, and the less chlorinated ones by aerobic microbes. Biotransformation rates, however, tend to be very slow such that the compounds are generally considered to be resistant to biodegradation.

Table 3.2 presents the 50 most frequently observed chemicals and the percentage of sites at which they were observed. The data are based on a survey of 888 proposed and final National Priorities List (NPL) sites.

3.11 SUMMARY

The United States is engaged in a massive effort to mitigate the effects of the environmental policies of the past 100 years. As a result, remediation site workers are required to work at sites contaminated with hundreds of hazardous chemicals, some with unknown physical properties and hazards. These chemicals can cause major injury to workers when exposure occurs. Therefore, a basic knowledge of the chemistry of the major potential contaminants at these sites is required so that workers can make educated decisions or evaluate the decisions of others in the selection of protective equipment and controls.

REVIEW QUESTIONS

1. What is the difference between molecular mass and molecular weight?
2. Explain how electrons are distributed in ionic and covalent compounds using the Lewis dot diagrams for NaOH and ethane.
3. In the following equations identify the oxidizing agent and reducing agent and balance the equations.

TABLE 3.2 ■ **The Fifty Most Frequently Observed Chemicals at NPL Sites**

Chemical Name	Frequency	Chemical Name	Frequency
Trichloroethylene	311	1,2-Dichloroethane	64
Lead	286	Chlorobenzene	64
Toluene	243	Carbon tetrachloride	61
Chromium and compounds	220	Heavy metals	56
		Pentachlorophenol	53
Benzene	208	Naphthalene	48
Chloroform	179	Methyl ethyl ketone	42
Polychlorinated biphenyls	159	Trichloroethane	38
		Iron and compounds	33
1,1,1-Trichloroethane	151	Barium	32
1,1,2,2-Tetrachloroethene	149	Volatile organics	31
Zinc and compounds	142	Manganese and compounds	31
Cadmium	141	Acetone	30
Arsenic	141	Phenanthrene	28
Phenol	121	Benzo(a)pyrene	27
Xylene	113	Chromium, hexavalent	27
Ethylbenzene	111	1,1,2-Trichlororthane	25
Copper and compounds	106	Arsenic and compounds	25
1,2-*trans*-Dichloroethylene	104	Dichloroethylene	24
		DDT	22
Methylene chloride	91	Styrene	22
1,1-Dichloroethane	85	Anthracene	22
1,1-Dichloroethene	79	Lindane	21
Mercury	78	Bis(2-ethylhexyl)phthalate	21
Cyanides (soluble salts)	73	Tetrachloroethane	21
Vinylchloride	70	Selenium	20
Nickel and compounds	65		

Source: CERCLA Site Discharges to POTWS: Treatability Manual (USEPA, 1990).

$$H_2 + Cl_2 \longrightarrow HCl$$

$$Fe + O_2 \longrightarrow Fe_2O_3$$

$$Mg + H_2SO_4 \longrightarrow MgSO_4 + H_2$$

4. How many grams of sodium chloride (NaCl) are present in 100 ml of a 2.45 M NaCl solution?
5. Draw the following compounds using a single line to indicate single carbon–carbon bonds and double lines to indicate double bonds:
   Methyl ethyl ketone
   1,2-Dichloroethylene
   Methyl alcohol
   Which of the above are unsaturated?

REFERENCES

Barbee, G. C., Fate of Chlorinated Aliphatic Hydrocarbons in the Vadose Zone and Ground Water, *Ground Water: Monitoring and Remediation*, Ground Water, Dublin, OH, Winter 1994.

Brady, J. E., and G. E. Humiston, *General Chemistry*, 4th ed., Wiley, New York, 1986.

Drago, R. S., *Principles of Chemistry with Practical Perspectives*, Allyn and Bacon, Boston, MA, 1974.

Ebbing, D. D., *General Chemistry*, 2nd ed., Houghton Mifflin, Boston, MA, 1987.

EPA, *CERCLA Site Discharges to POTWS: Treatability Manual*, U.S. Government Printing Office, Washington, D.C., 1990.

Henold, K. L., and F. Walmsley, *General Principles, Properties, and Reactions*, Addison-Wesley, Reading, MA, 1984.

Manahan, S. E., *Fundamentals of Environmental Chemistry*, Lewis, Ann Arbor, MI, 1993.

Manahan, S. E., *Hazardous Waste Chemistry, Toxicology and Treatment*, Lewis, Ann Arbor, MI, 1990.

INDUSTRIAL HYGIENE

CHAPTER OBJECTIVES

When you have completed this chapter, you will be better able to

- Define terms used in industrial hygiene and the chemistry of hazardous materials
- Better describe the possible effects of exposure to chemicals in the workplace
- Understand the basis for the workplace standards (PELs, TLVs)
- Understand the chemical and physical hazards of the workplace

4.1 TOXICOLOGY

Hazardous wastes contain many organic and inorganic compounds. The toxicities of these compounds to exposed workers and the environment are of utmost concern and the subject of intense study. This section summarizes the toxicological aspects of many constituents of hazardous wastes.

Toxicology is the science that deals with the effects of poisons upon living organisms. A poison or toxicant is a substance that, above a certain level of exposure or dose, has detrimental effects on tissues, organs, or biological processes.

Alcohol is a good example. Taken in small quantities, alcohol may be harmless and sometimes even medically recommended. However, an overdose causes intoxication and, in extreme cases, death. Similarly, vitamin A is required for normal functioning of most higher organisms, yet an overdose of it is highly toxic.

4.1.1 Definitions

Acute Effects: An acute health effect means that the body's response occurs at the time of exposure or within 24 hours. Acute effects may result from exposure to high concentrations of a substance. Examples of acute health effects include:

- Burns
- Choking
- Coughing
- Dizziness
- Nausea
- Death

Carcinogens: Agents that can cause cancer. A good example is benzene, which is a unwanted constitutent of gasoline. Other carcinogenic agents are halogenated hydrocarbons such as carbon tetrachloride and DBCP (dibromochloropropane).

Chronic Effects: A chronic health effect means that the body's response takes a long time, perhaps months or even years. Chronic effects involve repeated or prolonged exposure to a chemical. Examples of chronic effects include:

- Asbestosis
- Cancer
- Liver disease
- Lead poisoning

Hazard: The probability that harm will occur.

Hazardous Substance: A material with a substantial potential to pose a danger to living organisms, materials, structures, or the environment.

Hazardous Waste: A waste that may cause substantial damage to the health or environment when improperly managed.

Morbidity: Illness or impairment of function.

Mortality: Death.

Mutagenic: Capable of producing genetic changes or damage.

Nephrotoxic Agents: Chemicals that affect the kidney or nephros. Examples of these chemicals are cadmium, lead, and mercury.

Neurotoxins: Chemicals that directly affect the brain and/or the nerves, leading to neural toxicity. Examples of neurotoxins are chlorinated hydrocarbon pesticides (DDT, chlordane).

Risk: Risk is the probability that a substance will produce harm under specified conditions.

Risk Assessment: Risk assessment is the determination of the probability that an adverse effect will be produced.

Safety: Safety is the reciprocal of risk, the probability that harm will not occur under specified conditions.

Teratogenic: Capable of causing birth defects without adversely affecting the mother. Examples of teratogens are Thalidomide and dioxins.

Toxicity: Toxicity is the inherent ability of a material to cause damage to the body.

An example of the use of these definitions is as follows:

Potentially supertoxic substances can be used safely provided that one controls the environment to prevent the absorption of significant quantities of the material, thus avoiding toxicity. In such a situation, although the chemical is supertoxic, it is not hazardous in the manner in which it is being used. Therefore, depending on the conditions under which it is being used, a very toxic chemical may be less hazardous than a relatively nontoxic one.

The risk of injury would be low in this example because the potential for absorption of the supertoxic substance is low; that is, adequate controls are being used.

4.1.2 Chemical Classifications

Nuisance Materials These materials cause localized irritation to eyes, skin, mucous membranes, and respiratory tract. They cause no long-term or systemic effects. Examples of nuisance materials are calcium carbonate dust (the principal ingredient in limestone, marble, and chalk), calcium sulfate (the principal ingredient in gypsum), paper fiber, plaster of paris, and certain bulk particulate absorbant materials such as corn cobs. These inert dusts are relatively harmless unless exposure is severe.

The American Conference of Governmental Industrial Hygienists (ACGIH) no longer list these materials as a specific category.

Irritants. The toxic effects of these materials can include simple irritation, caused by an aggravation of whatever tissue the materials contact. Contact of some materials with the face and upper respiratory system affects the eyes and the tissues lining the nose and the mouth.

Many industrial chemicals are not truly toxic because they do not produce irreversible damage to some organ; nonetheless, at fairly low concentrations, they will irritate tissues with which they come in contact.

To a large extent the solubility of the irritant gas, vapor, mist, or particulate influences the affected part of the respiratory tract.

1. *Very soluble* chemicals cause upper respiratory tract irritation. Example of these are aldehydes, alkali dusts and mists, ammonia, acids, sulfur dioxides, and sulfur trioxides.
2. *Moderately soluble* chemicals can cause irritation of the upper and middle areas of lungs. Examples of these are halogens, ozone, isocyanates, and phosphorus halides.
3. *Insoluble* chemicals can cause irritations of the terminal respiratory passages and air sacs. Examples are nitrogen oxides, phosgene, arsenic halides.

Asphyxiants (Anoxia-Producing Materials) These chemicals prevent the uptake or use of oxygen by the body. The four general classes of asphyxiants are as follows:

1. *Simple Asphyxiants.* These displace oxygen and cause suffocation. Examples of these agents are carbon dioxide, ethane, nitrogen, helium, and methane.
2. *Anemic Asphyxiants.* These bind to hemoglobin and prevent oxygen transport. Examples of these agents are carbon monoxide, arsine, and analine.
3. *Histotoxic Asphyxiants.* These prevent oxygen utilization at the cellular level. An example of these agents is hydrogen cyanide.
4. *Direct-Acting Asphyxiants.* Paralysis of the respiratory center of the brain by hydrogen sulfide is an example.

Narcotics and Anesthetics These are chemicals that affect the central nervous system. Their principal effect is simple anesthesia without serious systemic effects, unless the dose is massive. Depending on the concentration, the depth of anesthesia ranges from mild symptoms to complete loss of consciousness and death.

A wide variety of hydrocarbon compounds cause narcotic or anesthetic effects. Some of these chemicals also cause other systemic effects. Anesthetics and narcotics include

- Aliphatic alcohols (ethyl, propyl, butyl, etc.)
- Aliphatic ketones (acetone, methyl ethyl ketone, etc.)
- Higher unsaturated hydrocarbons (acetylene, butylene, etc.)
- Ethers (ethyl, isopropyl, etc.)
- Paraffinic hydrocarbons (butane, etc.)
- Esters (produced by the conversion from alcohol in the body)

Fibrosis-Producing Materials Lung diseases are caused by the body's reaction to an accumulation of dust in the lungs. Fibrosis of the lung is a condition involving scar tissue formation in the lung. The restriction of lung function caused by this scarring places an additional burden on the right side of the heart, which tries to pump more blood to the lungs to maintain an adequate oxygen supply.

Fibrosis-producing materials include the following substances:

Silica. Silicon dioxide is a term used generally when referring to amorphous silica (noncrystalline), crystallized silica such as sand (quartz), and silicates such as clay (aluminum silicate). Only the crystalline (free silica) material found in quartz, tridymite, cristobalite, and a few other nonsilicate materials causes silicosis.

Silicosis can be classified as a lung disease caused by the inhalation of free silica dust. Silicosis is common in industries and occupations where the crystalline form of free silica dust is present, such as foundries, glass manufacturing plants, granite cutting operations, and mining and tunneling operations in quartz rock. Silicosis, which occurs throughout the world, has had many names, such as miner's asthma, grinder's consumption, miner's phthisis, potter's rot, and stonemason's disease. All these names, however, describe the same disease that is caused by dust from the crystalline form of free silica—usually quartz.

Asbestos. Asbestos causes asbestosis, another kind of lung disease that involves scarring of the lung and reduced lung function. This condition results from the inhalation of asbestos dust.

Asbestos fibers are generally characterized by high tensile strength, flexibility, and heat and chemical resistance. Certain grades of asbestos can be carded, spun, and woven, while others can be laid and pressed to form paper or used for structural reinforcement of minerals such as cement, plastic, and asphalt.

Asbestos in its commercial forms has been shown to be associated with the development of a variety of diseases. These include the following:

Asbestosis: A diffuse, interstitial, nonmalignant scarring of the lungs.

Bronchogenic Carcinoma: A malignancy of the lining of the lung's air passages.

Mesothelioma: A diffuse malignancy of the lining of the chest cavity (pleural mesothelioma) or the lining of the abdomen (peritoneal mesothelioma).

Cancer of the Stomach, Colon, and Rectum: The association between asbestos and these forms of cancer has not yet been well established.

Other lung diseases due to the inhalation of particulates include the following:

Berylliosis. Chronic beryllium disease results from inhalation of beryllium particulates. This disease is characterized by granulomas in the lung, skin, and other organs. Symptoms include cough, chest pain, and general weakness. Pulmonary dysfunction and systemic effects, such as heart enlargement, enlargements of the liver and spleen, cyanosis, and the appearance of kidney stones, also characterize the chronic illness. Beryllium has also been shown to cause cancer in rats and rhesus monkeys. This material has been classified as a suspect carcinogen in humans.

Coal Dust. Black lung disease is the name given to all lung diseases associated with chronic overexposure to coal dust. These diseases include chronic bronchitis, silicosis, and coal worker's pneumoconiosis.

Sensitizers Some primary skin irritants also sensitize. Certain irritants sensitize a person so that a dermatitis develops from a very low, nonirritating concentration of a compound that previously could have been handled without a problem. Initial skin contact with them may not produce irritation, but after repeated or extended exposure some people develop an allergic reaction termed *allergic contact dermatitis*.

Examples of substances that are irritants and allergens include turpentine, formaldehyde, chromic acid, amines, isocyanates, azo dyes, and epoxy resin components. Certain metals, such as nickel, chromium, and cobalt, are also sensitizers capable of triggering an allergic reaction.

Systemic Poisons These substances enter the body by inhalation, absorption, ingestion, or direct contact. They reach a specific target organ via the bloodstream, where they produce their effect.

Hepatotoxins: Chemicals that affect the liver. Examples of these chemicals are aromatic hydrocarbons, chlorinated hydrocarbons, and heavy metals.

Nephrotoxic Agents: Chemicals that affect the kidney or nephros. Examples of these chemicals are cadmium, lead, and mercury.

Neurotoxins: Chemicals that directly affect the brain and/or the nerves, leading to neural toxicity. Examples of neurotoxins are chlorinated hydrocarbon pesticides (DDT, chlordane).

Reproductive Toxins: Chemicals that affect the reproductive tissues. Examples of these chemicals are lead and dibromochloropropane (DBCP).

4.1.3 Entry Routes of Toxicants into the Body

The major routes of toxicant entry include absorption, ingestion, inhalation, and injection.

Toxic substances typically enter the body through the skin or eyes by absorption, through the mouth by ingestion (swallowing), through the lungs by inhalation (breathing), and through wounds by injection. Actually, all toxic chemicals must be absorbed into the body's cells to be poisonous. In the cases of ingestion and inhalation, the absorption takes place at a point away from the external body surface. Many toxic substances are able to enter the cells of the body in more than one of these ways, sometimes all four ways.

Inhalation Inhalation is the most common route for toxic substance absorption in the workplace. The large volume of air (and dust, fumes, vapors, mists, and gases) breathed in by workers—about 12,240 liters of air (432 cubic feet) each day—makes this route of entry particularly dangerous. The lungs are extremely vulnerable to chemical agents. Even substances that do not directly affect the lungs may pass through the lung tissue into the bloodstream.

Types of airborne contaminants are listed below:

Dusts: Solid particles produced by some mechanical action such as grinding, crushing, and so on.

Fumes: Very small, solid particles, usually less than 1 micrometer (μm), formed by condensation of heated volatile solids as they cool (welding materials).

Mists: Small, suspended liquid droplets produced by condensation of vapors or dispersion of liquids (may occur over plating baths).

Gases: Formless fluids that occupy the entire space in which they are confined.

Vapors: Gaseous form of materials that are normally liquids or solids at room temperature (produced by evaporation—for example, water converted to water vapor).

Inhalation of toxic substances results in a number of serious effects such as:

Direct or Local. Damage to the lungs occur. Asbestos, for instances, produces a scarring of the lung tissue and may result in a form of cancer called *mesothelioma*. Silica and coal dust also affect the lungs. Chemicals too can cause direct damage to the lung tissue. Inhalation of the mists produced by acid baths in plating operations can cause immediate damage and scarring of the lung.

Systemic. Damage occurs through absorption of the substance into the bloodstream and transport to the organs and tissues (brain, kidneys, liver, etc.). The lung contains little sacs, the alveoli, which have a very large surface area (about the size of a tennis court) for absorption. These alveoli are surrounded by tiny blood vessels or capillaries that enable the chemicals to pass into the body and reach their target organ(s) via the bloodstream.

Asphyxiants. These pose a special danger to workers, particularly in confined spaces. Air contains about 21% oxygen, the remainder being mostly nitrogen. Since a continuous supply of oxygen is necessary for life, a lack of oxygen results in asphyxiation or suffocation. In high concentrations, some substances (carbon dioxide, acetylene, methane, nitrogen, aliphatic hydrocarbons) reduce available oxygen in the air or the air, causing the asphyxiation. These gaseous substance are call *simple asphyxiants*. Some chemicals (carbon monoxide, hydrogen cyanide, aniline) combine with the hemoglobin in red blood cells, preventing the hemoglobin from from picking up oxygen and transporting it to the body's cells. These chemical asphyxiants are toxic because they indirectly result in asphyxiation.

Skin Exposure The skin is a barrier between the environment and the cells of the body. Skin diseases are the most common occupational illness. This is not surprising, since workers frequently get chemicals on the skin.

There are four possible effects of skin exposure:

- No effect, because the skin acts as an effective barrier to the chemical
- The skin becomes irritated
- The skin and person become sensitized
- The chemical passes through the skin, enters the bloodstream, and migrates to its target organ(s)

Exposure may result from skin contact with solids, liquids, and gases. Important responses to skin exposure are described below:

Corrosion. This is the eating away of skin tissue by strong chemicals like acids and caustics.

Contact Dermatitis. This results from direct chemical contact and appears at the site of contact. Most common on the hands, it usually goes away after exposure to the irritant is stopped. Some chemicals causing dermatitis include dilute acids and caustics, formaldehyde, ammonia, turpentine, metal dusts, and organic solvents.

Allergic Contact Dermatitis. This occurs when an individual becomes sensitized by prior exposure to a toxic substance. Poison ivy is a good example of this allergic response. A later exposure produces symptoms 2–3 days after exposure. Examples of chemicals frequently involved in allergic contact dermatitis include sodium bichromate, epoxies, aromatic amines, formaldehyde, and nickel metal.

Acne. Pimples can appear at sites of exposure. A familiar form of this response, called *chloracne*, results from exposure to chlorine. Known acne-producing agents include chlorine, oil, and tar.

Skin Cancers. Tumors of the skin may take 20–30 years to occur after toxic exposure. They may be caused by mineral oils, tars, and arsenic.

Symptoms that may result from chemical exposure include:

- Acne
- Blisters
- Discolored patches, bumps, or moles
- Itching
- Rashes or redness
- Swelling
- Ulcers, cracks

Eye Exposure In general the eyes are affected by the same chemicals that affect skin. However, the eyes are more sensitive than skin. Examples of chemicals with special toxicity for eyes include formaldehyde, ammonia, and chlorine gas.

Ingestion The oral route, usually a minor pathway for workplace exposures, should be easily eliminated through strict observance of hygienic practices relating to eating, drinking, and smoking. It may also result from involuntarily swallowing mucus from the respiratory system.

The effects of ingestion are not necessarily limited to the direct contact with the tissues of the digestive system. Ingestion can affect other target organs by the absorption of the chemical into the bloodstream.

Injection Chemical exposure by injection must be prevented. Chemicals can be introduced into the body through puncture wounds by stepping, tripping, grabbing, and falling onto contaminated sharp objects. Protective gloves and safety shoes or boots with steel shanks are important protective measures against injection.

4.1.4 Fate of Chemicals That Enter the Bloodstream

The human body deals with toxic chemicals in three main ways: metabolism, excretion, and storage.

Metabolism Chemicals absorbed into the bloodstream from the stomach, intestine, or lung make their way to the body's largest organ, the liver. There useful nutrients like glucose are stored, sent to other organs, or converted to other useful chemical compounds. This process of converting one chemical into another is called *metabolism*; toxic compounds undergo metabolism too. Two main things may happen to a toxic compound as it is metabolized in the liver:

1. *Detoxification.* The toxic substance is converted into a harmless substance, Sometimes the metabolite is water-soluble and can be excreted from the body. The liver is the body's main detoxifying organ.
2. *Formation of Reactive Intermediates.* Sometimes the liver converts toxic chemicals to more toxic ones. These then leave the liver and blood and make their way to the various tissues of the body, where they can cause damage.

Excretion The toxic chemical is given back to the environment in the air breathed out of the lungs or in the urine formed by the kidneys. Often the body is not able to excrete all the substance quickly, so some remains in the tissues for a long time. Some toxic chemicals cannot be excreted.

Storage Many chemicals, particularly those that are soluble in oil or fat, are not removed from the body but are stored instead. Usually this occurs in the fat tissue. Examples of chemicals stored in this fashion are PCBs, DDT, and dioxins.

4.1.5 Toxic Chemicals and Target Tissues

Certain toxic chemicals have effects on special cells of the body. For example, doctors in the 19th century discovered that chimney sweeps exposed to chimney soot had a very high incidence of scrotal cancer. We now know that the organic chemical residues from chimney soot are very potent cancer-producing agents. More recently, a number of halogenated pesticides have been linked to adverse responses of the male reproductive system. Factory workers exposed to the compound DBCP (dibromochloropropane) became sterile or exhibited low sperm counts. Other halogenated pesticides that have elicited similar toxic responses in men are kepone and DDT.

Organophosphate insecticides (parathion, malathion, diazinon, etc.) and carbamate insecticides (carbaryl or sevin, aldicarb, etc.), both used in agriculture directly affect nerve cells. Tremors and spasms result.

The kidney is the organ most sensitive to the toxic effects of cadmium. In factories where nickel/cadmium batteries are manufactured, workers who are exposed to excessive amounts of cadmium oxide exhibit consistent indications of renal toxicity. Mercury and lead can also cause renal damage. These chemicals—cadmium, lead, and mercury—are known as *nephrotoxic agents*.

Measurement of Toxicity Toxicity measurements rely on the following nomenclature:

Lethal Dose (LD): Amount of administered material required to cause the death of a test population (LD_{100})

Lethal dose 50 (LD_{50}): Amount of administered material required to kill one-half the test population

Lethal concentration (LC): Inhalation exposures are expressed as LC rather than LD

Lethal concentration 50 (LC_{50}): Amount of exposure (material inhaled) required to kill one-half the test population

Dose–Response Relationship Generally, a given amount of toxic agent will elicit a given type and intensity of response. This dose–response relationship is the basis for measurement of the relative harmfulness of a chemical. Because humans cannot be used as test organisms, almost all toxicological data are derived from other mammalian species, with the results extrapolated to humans. The test organism is chosen for its ability to simulate human response. For example, most skin tests are performed on rabbits because their skin response most closely resembles that of humans.

In much of the toxicological testing, the response measured is death. The dose is the amount of chemical administered to the animal. A typical dose, for example, would be expressed in milligrams or micrograms of test agent per kilogram of body weight. The test data are plotted as a dose–response curve as shown in Figure 4.1. From this curve, the dose that killed a certain percentage of test organisms can be calculated. This dose is called the *lethal dose*. Most often, experiments are designed to measure the dose that kills 50% of the test organisms. This is the lethal dose 50, or LD_{50}, and is a relative measurement of toxicity. If a compound A has an LD_{50} of 100 mg/kg and compound B has an LD_{50} of 500 mg/kg, compound A is

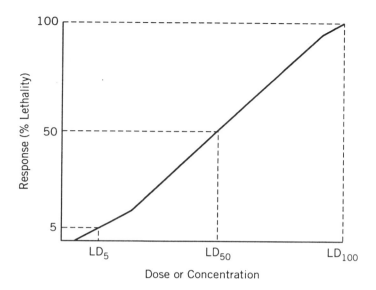

FIGURE 4.1 ▪ Dose–response curve illustrating the percent response of the test group at various doses.

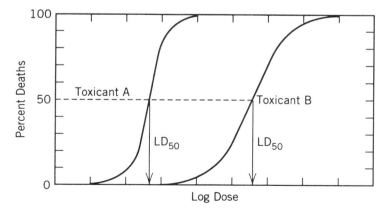

FIGURE 4.2 ▪ Dose–response curve showing two chemicals of differing toxicity.

more toxic than compound B because A gives the same response (50% deaths) as B at a lower dose (see Figure 4.2). A value similar to the LD_{50} is the lethal concentration 50, or LC_{50}, used for inhalation exposures. The value is measured as parts per million of toxic agent per exposure time (ppm/hr).

Another important factor to consider when determining the toxicity of a material is the relationship between concentration and exposure time. Generally, an acute exposure refers to a large single dose received over a short period of time. The difference (in terms of deleterious effects) is that a small, acute exposure may result in no effect on an organism, while a chronic exposure to the same dose may be harmful. On the other hand, a large, single dose in a short period of time might be much more hazardous than the same dose administered over a longer time.

Possible responses elicited are as follows:

- *Adjustment.* Body adjusts and maintains normal balance—no residual effect.
- *Compensation.* Body compensates for introduction of chemical—effect is reversible (damage repaired).
- *Impairment.* Some level of damage that is irreversible (may not be noticeable).
- *Disability.* Permanent damage which is noticeable and causes some loss of function.
- *Death.* May be from acute or chronic exposure.

4.1.6 Basis for Workplace Standards

Chemical Analogy When evaluating a new chemical, researchers often do not have animal or human toxicity data. Therefore, the researcher may assume that the nature of the response to the chemical will be similar to that produced by exposure to a substance with a similar chemical structure.

Animal Studies The toxic effects of chemicals should be known before they are introduced into the workplace. Preventive measures can then be designated to protect workers. For new chemicals with little or no data available, animal studies can be an important method of developing this new information.

The toxicological effects of such chemicals can be determined in the lab by exposing animals to known concentrations for controlled periods of time. This is followed by more definitive studies involving several species of animals. The researchers quantify the exposure in terms of concentration and duration. The results of these studies are extrapolated to humans and used to set safe or acceptable limits for exposure in the workplace.

Human Epidemiological Studies These studies, which attempt to identify relationships between diseases and occupational exposures, compare the number of affected people in one group to number of affected people in another group. These types of studies can identify any significantly higher incidences of a disease. If researcher observe a higher incidence in a given population, a search for the reason begins. Sometimes the reason can be an exposure to a chemical in the workplace, contamination of drinking water supplies, the genetic makeup of the population, and so on.

Practical Considerations The most immediate concern is the protection of site workers and the environment. The following factors must be considered:

- What toxic agents are present?
- How much of the agent is present?
- How will it enter the body?
- How will it affect the body?

Answers to these and related questions will dictate how to protect personnel (types of respiratory and protective gear needed) and what monitoring to require (e.g., continuous or intermittent, direct reading or grab sample).

4.1.7 Threshold Limit Values (TLVs)

These values, which refer to airborne concentrations of substances, represent conditions under which nearly all workers may be repeatedly exposed day after day without adverse effect.

Because of the wide variation in individual susceptibility, a small percentage of workers may experience discomfort from some substances at concentrations at or below the threshold limit value.

A smaller percentage of workers may be affected more seriously by aggravation of a preexisting condition or by development of an occupational illness.

Hypersensitive individuals may not be adequately protected.

How Are TLVs Established? The TLVs are recommendations of the American Conference of Governmental Industrial Hygienists (ACGIH). Each year industrial hygienists from government and academia meet to review data.

The TLVs are based on the best available information from industrial experience, from experimental human and animal studies, and, when possible, from a combination of these.

In 1946, there were TLVs for 148 substances. In 1968, OSHA adopted into their standard the ACGIH–TLV guidelines. And in 1989, OSHA updated their standards by adopting the 1988 TLVs (currently these updated standards have been suspended pending further adjudication). These can be found in Title 29 CFR 1910.1000—Permissible Exposure Limits (PELs). Today, there are approximately 800 TLVs, and TLV recommendations for new substances are added each year as toxicological and industrial data become available.

Documentation The policy of the TLV committee of the ACGIH is to prepare a justification for each proposed TLV. From time to time these are published in a document entitled *Documentation of Threshold Limit Values*, available from the ACGIH. This document should be consulted whenever a particular TLV is to be applied.

The latest recommendations of the TLV committee are published each year by the ACGIH in the TLV book. This publication is a good starting point for reviewing potential health effects of materials of concern. However, the book will not list all materials encountered in the workplace.

Definitions

Threshold Limit Value—Time-Weighted Average (TLV-TWA): That concentration for a normal 8-hour workday or 40-hour workweek to which all workers may be repeatedly exposed, day after day, without adverse effect.

Threshold Limit Value—Short-Term Exposure Limit (TLV-STEL): A 15-minute, time-weighted average exposure that should not be exceeded at any time during a workday even if the 8-hour time-weighted average is within the TLV. Exposures at the STEL should be no longer than 15 minutes and should not be repeated more than four times per day. In addition, at least 60 minutes should intervene between successive exposures at the STEL. For those substances without a recommended STEL, excursions above the TLV-TWA should be controlled even where the 8-hour TWA is within recommended limits. The recommended short-term exposure limits should not exceed three times the TLV-TWA for more than 30 minutes during the workday; under no circumstances should they exceed five times the TLV-TWA, provided that the TLV-TWA has not already been exceeded.

Threshold Limit Value—Ceiling (TLV-C): The concentration that should not be exceeded even momentarily. Some materials produce intolerable irritation above certain levels. Others produce irreversible physiological effects. For these materials a ceiling exposure limit has been set. This is a "not to be exceeded at any time" concentration. The chemicals for which a ceiling limit has been established are denoted by a "C" in the TLV book and the OSHA standards.

Permissible Exposure Limits (PELs): These limits, set forth in the regulations promulgated by OSHA, are therefore the legally enforceable exposure limits. They can be changed only by full congressional rule-making procedures. Thus, they are seldom updated. In 1989, OSHA adopted the 1988 TLVs, which are codified in Tables Z-1, Z-2, and Z-3 of 29 CFR 1910.1000.

"Skin" Notation: More than 25% of the substances listed in the TLV book are followed by the designation "skin." This refers to the potential for skin, eye, and/or mucous membrane absorption caused by direct contact with the chemical. The amount absorbed can be significant, sometimes exceeding the dose received through inhalation and ingestion. The possibility of skin absorption must be taken into consideration when working with or near these chemicals. Special personal protective equipment may be required.

Limitations of TLVs TLVs, intended for use in the practice of industrial hygiene, are guidelines or recommendations for the control of potential health hazards. They are not intended for use as

- Indices of, or limits for, community air pollution
- A relative index of toxicity
- Limits for exposures encountered in extended work shifts (greater than 8 hours per day 40 hours per week)
- Proof of hazard

- Guidelines in countries with work conditions differing from those in the United States

Special Consideration—Mixtures Chemical materials do not exert their toxic effects independently of each other. The workplace may contain two or more hazardous substances that act upon the same target organ. In cases such as these, the primary consideration should be their combined effect, rather than that of either component. In the absence of information to the contrary, the effects of the different chemical hazards should be considered as additive.

Synergistic action or potentiation may occur with some combinations of contaminants; therefore, different combinations or mixtures should be evaluated individually. Special formulas are available to determine if the total exposure for similarly acting chemicals has been exceeded. Appendix C of the TLV book covers the evaluation of mixtures.

Biological Exposure Indices (BEIs) Biological monitoring provides occupational health personnel with a tool for assessing workers' exposure to chemicals. The monitoring and measurement of workplace ambient air provides an assessment of worker inhalation exposure to chemicals. The TLVs serve here as reference values. Biological monitoring consists of an assessment of overall exposure to chemicals that are present in the workplace; the process involves measurement of appropriate determinants in biological specimens collected from the worker at the specified time. The BEIs serve as a reference value.

The determinant can be the chemical itself or its metabolite(s), or a characteristic reversible biochemical change induced by the chemical. The measurement can be made in exhaled air, urine, blood, or other biological specimens collected from the exposed worker.

Biological exposure indices are reference values intended as guidelines for the evaluation of potential health hazards in the practice of industrial hygiene. The BEIs represent the levels of determinants that are likely to be observed in the specimens collected from a healthy worker who has been exposed to a chemical to the same extent as a worker with inhalation exposure to the TLV.

These values apply to 8-hour workday exposures, 5 days a week. Biological exposure indices for altered working schedules can be determined by extrapolation.

Examples of Determinants Analysis of urine tests for the following compounds:

Arsenic	Mercury	Selenium
Benzene	Nickel	Thallium
Cadmium	Nitrobenzene	Uranium
Cyanide	Parathion	Zinc

Analysis of blood samples tests for the following compounds:

Aluminum	Carbon monoxide	Mercury
Cadmium dust	Lead	Methyl bromide

Analysis of exhaled air also tests for the following compounds:

Alcohols

Aliphatic hydrocarbons

Chlorohydrocarbons

Ketones

Examples of metabolite or indirect analyses are:

Product in Urine	**Toxic Agent**
Formic acid	Methyl alcohol
Hippuric acid	Ethyl benzene
	Styrene
	Toluene
Methylhippuric acid	Xylene
p-Nitrophenol	Parathion
Phenol	Benzene
Thiocyanate	Cyanate

Health workers can also analyze blood for specific biochemical constituents such as methemoglobin, which is an indicator of exposure to nitrites.

Limitations of BEIs Because of the wide variation in human biochemical characteristics, BEIs should be used with caution.

Worker baseline or normal (nonexposed) data should be available for comparison to exposed levels. Factors that can affect baseline values are

Individual health status

Individual lifestyle—that is, diet, hobbies, and so on

Environmental sources

Carefully read the limitations in the TLV book before using these indices.

Notice of Intended Changes Proposals for adding new chemicals to the TLV book or for recommending changes in the adopted values are published in the "changes" section of the book. In both cases, the proposed values should be considered as limits that will remain in this listing for at least one year. If after one year no evidence questions the appropriateness of these values, the values will be added to the list.

Other Materials in the TLV Book

Temperature stress—heat and cold stress

Noise—continuous and impact noise

Microwave and RF radiation

Ultraviolet radiation

Hand–arm vibration

Carcinogens

Chemical and physical agents under study for inclusion in the future

4.2 CHEMICAL HAZARDS

4.2.1 Hazard Communication Standard

A hazardous chemical is any chemical or material used in the workplace that is regulated under OSHA's Hazard Communication Standard. A hazardous chemical is also defined as a physical or health hazard.

Chemical and Physical Hazards
- Combustible or ignitable liquids
- Compressed gases
- Explosive materials
- Flammable materials
- Organic peroxides
- Pyrophoric materials
- Unstable or reactive materials
- Water-reactive materials

Combustible or Ignitable Liquid The U.S. Department of Transportation (DOT) defines a combustible liquid as having a flash point equal to or greater than 100°F but less than 200°F.

The Environmental Protection Agency (EPA) defines an ignitable liquid as having a flash point at or below 140°F.

Flammable Materials

Aerosol: Projects a flame exceeding 18 inches at full valve opening

Gas: Forms a flammable mixture in air at or below 13%

Liquid: Flash point less than 100°F

Solid: Capable of causing a fire through friction, absorption of moisture, spontaneous chemical change, or retained heat

Compressed Gas This is a gas whose pressure is greater than 40 psi (pounds per square inch) at 70°F or whose pressure is greater than 104 psi at 130°F, or a liquid having a vapor pressure greater than 40 psi at 100°F.

Explosive This is a material capable of causing a sudden, almost instantaneous release of pressure, gas, and heat when subjected to sudden shock, pressure, or high temperature.

Organic Peroxides These are organic compounds with a —O—O— structure. The chemicals are strong oxidizing agents that may be serious fire and explosion hazards. They are typically sensitive to heat and shock. Some representative organic peroxides are methyl ethyl ketone peroxide, benzoyl peroxide, and peracetic acid. Note that, as with inorganic peroxides, the prefix "per" often occurs in the chemical name.

Oxidizing Agents These are materials that initiate or promote combustion, causing fire by the release of oxygen. Examples are perchlorate, peroxides, fluorine, ozone, metallic chlorates, and permanganate. Addition of oxidizers to fuels will cause vigorous combustion (e.g., hypochlorite to brake fluid or gasoline). Therefore, fires or explosions are more likely to occur if oxidizers are stored near flammables.

Pyrophoric Materials These are materials that ignite spontaneously in air. Some representative pyrophoric materials are white phosphorus, the alkali metals (i.e., lithium, sodium, potassium, etc.), and powered forms of magnesium, calcium, cobalt, manganese, iron, zirconium, and aluminum. Also included are some organometallic compounds (such as lithium ethyl and lithium phenyl), and some metal carbonyl compounds (such as iron pentacarbonyl). Another class of pyrophoric compounds consists of metal and metalloid hydrides, including lithium hydride, pentaborane, and arsine. Moisture in air is often a factor in spontaneous ignition.

Unstable Materials These are materials that polymerize, decompose, condense, or become self-reactive under conditions of shock, pressure, or elevated temperature. An example of an unstable material is picric acid.

Water-Reactive Material These materials react violently with water to release a gas that is either flammable or a health hazard. An example is the addition of lithium, sodium, potassium, or other members of the alkali metals to water. This causes a strong reaction, releasing hydrogen and heat. The heat produced is great enough to cause the hydrogen to react explosively with the oxygen in air to form water. A similar reaction occurs when hydrides, such as lithium hydride, react with water.

4.2.2 Chemical and Physical Characteristics

Understanding how chemicals behave can help the site worker, plant employee, or first responder anticipate the hazards of the chemicals at the jobsite. Scientists have developed many terms to describe the chemical and physical properties of substances. Memorizing these terms is unimportant, but a general understanding of their meanings may be helpful.

Boiling Point This is the temperature at which a liquid changes into a gas. Also the temperature at which the pressure of vapor escaping from the liquid equals the outside pressure. This determines whether a substance will be a liquid or gas at ambient temperature. Chemicals with low boiling points are more dangerous than those with high boiling points. The boiling point is the lowest temperature at which maximum vaporization is occurring. Example: Water boils at 212°F, benzene at 176°F, ethyl alcohol at 173°F, and ethyl ether at 95°F.

Corrosion This is a process of material degradation. Upon contact, a corrosive material may destroy body, metals, plastics, and other materials. A corrosive agent is a reactive compound or element that produces a destructive chemical change in the material upon which it is acting.

Strong corrosives, such as hydrochloric acid or sulfuric acid, should be stored in glass or special plastic containers. If they are kept in metal containers, they will rapidly dissolve or digest the container. A strong base, such as sodium hydroxide, could also have the same effect. Examples of corrosives include nitric acid, phenol, and ammonia.

Corrosives can damage skin, lungs, eyes, mouth, and stomach if they come in contact with the body. The use of safe practices, including the use of chemical protective clothing and respiratory protection, may be required.

Explosive Range This is the range of chemical vapor concentrations in air that can be ignited. Gasoline's range is 1.4% to 7.6% vapor in air.

Lower Explosive or Flammable Limit This is the minimum concentration of vapor in air that will ignite. For gasoline this lower limit (LEL) is 1.4%.

Upper Explosive or Flammable Limit This is the maximum concentration of vapor in air that will ignite. For gasoline this upper limit (UEL) is 7.6%.

The explosive range of a flammable liquid and its lower explosive limit are perhaps the two most important characteristics to check when trying to determine a substance's flammability.

Substances with a low LEL and a wide explosive range must be considered dangerous fire or explosion hazards.

Flammability This is the potential for a liquid or gas to catch on fire. The flammability of a substance is important to know during site remediation and cleanup or during an emergency, where responders could determine the probability of a fire by knowing its flammability.

Flash Point This is the lowest temperature at which a flammable liquid gives off sufficient vapor in air to ignite.

The flash point is used to classify the relative fire hazards of liquids. If the flash point of a liquid is low, it is considered flammable and care should be exercised in its handling.

Examples of chemicals with low flash points are gasoline, which has a flash point of −45°F, and kerosene, which has a flash point of 43–72°F. Gasoline is clearly far more hazardous and more likely to cause handling problems on the job site.

pH pH is a measurement scale used to determine whether a substance is an acid or a base. The scale is a series of numbers ranging from 1 to 14. A pH of 7 is considered neutral. The pH range of 1–6 is acidic; the lower the number below 7, the more acidic the solution. Chemicals with a pH ranging from 8 to 14 are alkaline or basic. The higher the pH above 7, the more alkaline the chemical.

Sodium hydroxide, or lye, is an example of a base with a pH between 12 and 13. An acidic chemical is hydrochloric acid, which has a pH between 2 and 3. The pH of a chemical is important because mixing strong acids with strong bases could result in a violent reaction. In addition, acids and bases individually can cause skin burns or corrode certain metals and containers. The pH can be measured by litmus or hydrion paper or by a pH meter.

Physical State This consists of the physical characteristics of a substance at a given temperature, pressure, and volume. Substances can exist as solids, liquids, or gases. The physical state of a substance can determine the type of response, the required remediation and/or disposal activity, and the most appropriate protective equipment and clothing.

Solubility This is the measurement of how readily a chemical dissolves in liquid. The solubility of methyl bromide, for instance, is 2% (2 g/100 ml) in water at 68°F. Ethyl or methyl alcohol are miscible in water; that is, they readily mix with water in any ratio without separation into two phases. However, motor oil does not mix with water. Instead, it floats on the water's surface as a separate phase.

The importance of solubility becomes clear when a liquid spills into a waterway or pond. If the liquid is not soluble, it will either float to the top or sink to the bottom. To ascertain the location of the spilled material, you would have to determine the specific gravity of the material.

Specific Gravity The specific gravity of a material is the ratio of the weight of the material compared to an equal volume of water. This is expressed as a relative density to water.

Insoluble materials with specific gravities of less than 1.0 will float in water, while materials with specific gravities of more than 1.0 will sink in water. Most flammable liquids have specific gravities of less than 1.0; therefore, they will float on water.

Vapor Density This is the relative density or weight of a vapor or gas as compared to that of an equal volume of air. Air has a vapor density of 1. This is also the ratio of the weight of a volume of pure vapor or gas to the weight of an equal volume of dry air at the same temperature and pressure (heavier or lighter than air).

Helium, a gas, has a vapor density of <1, which means that it is lighter than air and will rise. A gas such as phosgene, which has a vapor density of >1, is heavier than air and will sink. This physical characteristic is important because it allows workers to predict the location of gas and vapors in confined spaces. If a vapor is denser than air, it may accumulate in low areas, such as the bottom of a sump or trench. Since the vapor can collect in low-lying areas, it could lead to a decrease in the amount of oxygen in the air, a fire, an explosion, or other dangers to an emergency responder, site worker, or maintenance person.

Vapor Pressure This is the pressure exerted by a vapor at a given temperature. The higher the vapor pressure, the more potentially hazardous the chemical. Vapor pressure is measured in millimeters (mm) of mercury or in pounds per square inch absolute.

Examples of vapor pressure are given in Table 4.1. Materials with high vapor pressure will vaporize rapidly and enter the air. Toxic chemicals can overcome a first responder or site worker who is not protected. Flammable chemicals with moderately high or high vapor pressures are extremely hazardous because of the potential for explosion. When working with these chemicals, workers must take care to prevent the accumulation of vapors in confined spaces.

TABLE 4.1 ■ Vapor Pressure of Several Common Chemicals

Chemical	Relative Pressure	Vapor Pressure (mm Hg)
Acetic acid (vinegar)	Low	11
Iodine	Low	0.3
1,1,2-Trichloroethane	Low	19
Benzene	Moderate	75
Hexane	Moderate	124
Ethyl ether	High	442
Pentane	High	400

Viscosity This is the internal resistance of a substance to flow. It is also the thickness of a liquid. Temperature affects viscosity. Water has a low viscosity, while motor oil has a greater viscosity.

Since viscosity can change with temperature, a high-viscosity substance may become very fluid when exposed to heat. This results in the substance's spreading faster and further.

Volatility This is the ability of a liquid to evaporate or vaporize. Gasoline is a volatile liquid. Volatile liquids can give off vapors that may be harmful to your health or flammable.

4.2.3 Health Hazards

Carcinogens These are cancer-causing substances. Cancer in humans is more difficult to predict than some other kinds of effects of exposure to toxic materials. The reasons for this are twofold:

- The vast majority of chemicals in use have not been tested for cancer-causing effects on laboratory animals. Of approximately 60,000 chemicals in use, only about 300 have been tested. Of these, about half have been shown to cause cancer in lab animals. Only 21 of these cancer-causing chemicals are regulated by OSHA.
- The results of exposure to a cancer-causing chemical can take 10–40 years to appear. You may be healthy for 20 years and get the cancer the very next year. The time it takes for the cancer to show up is called the *latency period*. Table 4.2 lists the latency periods of several carcinogens.

It may be too late to do anything about previous exposures, but any reduction in exposures now and in the future will reduce the risk of getting cancer.

Highly Toxic Agents These substances are defined by the following:

- Oral LD_{50} less than or equal to 50 mg/kg of body weight when tested in albino rats
- Dermal LD_{50} less than or equal to 200 mg/kg body weight when tested in albino rats
- LC_{50} less than or equal to 200 ppm for gases and vapors or 2 mg/liter for particulates.

TABLE 4.2 ■ Latency Periods of Several Common Carcinogens

Carcinogen	Average Latency Period (years)
Acrylonitrile	23
Arsenic	25
Asbestos fibers	30
Benzene	10
Benzidine and its salts	16
Chromium and chromates	21
Mustard gas	17
Naphthylamine	22
Vinyl chloride	15

Toxic Agents These substances are defined by the following:

- Oral LD_{50}, greater than 50 mg/kg but less than or equal to 500 mg/kg
- Dermal LD_{50}, greater than 200 mg/kg but less than or equal to 1000 mg/kg
- LC_{50}, greater than 200 ppm but less than or equal to 2000 ppm, or greater than 2 mg/liter but less than or equal to 20 mg/liter.

Irritants These noncorrosive materials have a reversible inflammatory effect on living tissue, caused by chemical action at the site of contact.

Corrosives Chemical action at the site of contact causes visible destruction of, or irreversible alterations in, living tissue.

Sensitizers These are materials that cause a substantial proportion of exposed people or animals to develop, after repeated exposure, an allergic reaction in otherwise normal tissue.

Hepatotoxins These are chemicals that affect the liver. Examples of these chemicals are aromatic hydrocarbons, chlorinated hydrocarbons, and heavy metals.

Nephrotoxic Agents These are chemicals that affect the kidney or nephros. Examples of these chemicals are cadmium, lead, and mercury.

Neurotoxins These are chemicals that directly affect the brain and/or the nerves, leading to neural toxicity. Examples of neurotoxins are chlorinated hydrocarbon pesticides (DDT, chlordane).

Reproductive Toxins These are chemicals that affect the reproductive tissues. Examples of these chemicals are lead and dibromochloropropane (DCBP).

Hematotoxins These materials affect the blood and blood-forming systems. Examples of these agents follow.

Chemical Asphyxiants
1. Carbon monoxide is the best-known chemical agent that decreases the oxygen transport of the blood and produces an anoxic or asphyxial hypoxia (both mean inadequate supply of oxygen). Carbon monoxide (CO) elicits this toxic effect because CO has a stronger binding affinity for hemoglobin than does oxygen. This means that when CO is present, it will preferentially be bound to hemoglobin rather than oxygen, causing this anoxic effect.
2. Cyanide poisoning causes cytoxic hypoxia, which results from an interference in the utilization of oxygen during cell metabolism, despite the presence of an adequate supply of oxygen and blood flow. In this situation, tissue oxygen concentration may be normal or higher than normal because the cyanide prevents oxygen use during continued oxygen uptake by the body. The lethal dose for an adult is approximately 100 mg of sodium or potassium cyanide. Death because of central respiratory arrest is rarely delayed more than an hour.
3. Hydrogen sulfide (HS^-) poisoning also causes cytoxic hypoxia. Hydrogen sulfide is common to sewers, landfills, and oil fields. The symptoms of HS^- poisoning are similar to those of cyanide poisoning, except that hydrogen sulfide is an irritant gas that may produce conjunctivitis of the eyes or pulmonary edema.

4.2.4 Target Organ Effects

Reproductive System

Lead and Other Metal Compounds. Occupational exposure to lead and other metal compounds decreases the fertility of male workers and increases the rate of spontaneous abortions in their wives. As long ago as the early 1900s, marital life records of men working in storage battery plants revealed that 24.7% had sterile marriages, compared to 14.8% in a control group. Another study indicated that wives of printers exposed to lead had an abortion rate of 14%, in comparison to 4% for the community.

In a 1983 study, blood lead-level concentrations as low as 53 μg lead/100 ml blood were associated with reduced fertility. In another recent study, blood lead levels as low as 41 μg/ml disclosed a significant increase in the number of abnormal sperm and a significant decrease in sperm count in men exposed to high occupational lead levels.

Cadmium, nickel, and methyl mercury are also male reproductive toxins because they cause testicular damage in men or animals.

Halogenated Pesticides. Chronic exposure to a number of halogenated pesticides has been linked to adverse responses of the male reproductive system. Factory workers who were exposed to the nematocide 1,2-dibromo-3-chloropropane (DBCP) became sterile. Some men were sterile because of low sperm counts; others were sterile because the germ cell population was completely destroyed.

Other halogenated pesticides that have elicited toxic responses in men are kepone and DDT.

Organic Solvents. Swedish workers who handle the organic solvents toluene, benzene, and xylene have been reported to have low sperm counts, abnormal sperm, and varying degrees of fertility.

Workers exposed to dinitrotoluene (DNT), used in the synthesis of toluenediamine (TDA), have been found to have reduced sperm counts, and several showed increased numbers of sperm with abnormal morphology.

Liver The effects of the *halogenated hydrocarbons* vary considerably with the number and type of halogen atoms present in the molecule. *Carbon tetrachloride* is highly toxic, acting acutely by injury to the kidneys, liver, central nervous system, and gastrointestinal tract. The current TLV for carbon tetrachloride is 5 ppm; however, NIOSH recommends that the permissible exposure limit (PEL) be reduced to 2 ppm and that it be regulated as an occupational carcinogen.

Chronic exposure to carbon tetrachloride also damages the liver and kidneys and is suspected of causing liver cancer. Table 4.3 describes many of the organic solvents in common use and their important chemical, physical, and physiological properties.

Nitrosamines such as *dimethyl nitrosamine* have been shown to cause cancer. Dimethyl nitrosamine was widely used as an industrial solvent. Studies of exposed workers revealed that this chemical caused liver damage and liver cancer. Different nitrosamines cause cancer in different organs.

Kidneys *Halogenated hydrocarbons* such as *carbon tetrachloride* and *chloroform* are nephrotoxins. These chemicals are activated by the cellular metabolism of the kidney to a toxic form that causes the damage.

Bromobenzene, tetrachloroethylene, and *1,1,2-trichloroethylene* also produce toxic effects to the kidney similar to those caused by carbon tetrachloride and chloroform.

Metallic mercury and its derivatives are widely used as catalysts and fungicides and for numerous industrial applications. The high volatility of mercury and its derivatives makes them especially dangerous, because they may enter the circulatory system via the respiratory route. Kidney cell damage from mercury and its derivatives is well documented.

Cadmium, which is a pulmonary toxin, is also a nephrotoxin. About 15–30% of inhaled cadmium is absorbed into the circulation from the respiratory system. Cadmium causes cellular damage to the kidney, thus effecting its function.

Chromium, another respiratory toxin, is also a nephrotoxin. Hexavalent chromium, as in chromate and bichromate, can cause cell damage and are confirmed human carcinogens.

Lead is another nephrotoxin. Because lead has numerous industrial applications, industrial workers can often be overexposed if proper safety equipment and engineering controls are not utilized. Lead also causes kidney cell damage, which can cause improper functioning of the kidneys.

TABLE 4.3 ■ Solvents in Common Use

Category	Class Toxicity	Relative Flammability[a]	Relativity Volatility[b]	Common Industrial Uses	Examples	Signs
Aromatic hydrocarbons	Defat skin—dermatitis Inhalation—narcotic, irritant; chronic blood changes; some CNS depressants	Variable—moderate to high	Mostly high	Chemicals, drugs, rubber manufacturing, paints, plastics, explosives	Benzene, toluene, xylene, naphthalene	Distinctive odors, but olfactory fatigue limits; eye and respiratory irritant. Benzene skin hemorrhages
Halogenated aromatic hydrocarbons	CNS depressant (narcotic). Liver, kidney; eye irritation; skin absorption	Variable—low to high	Mostly high	Solvent, chemical intermediate, insecticides, disinfectants	Chlorobenzenes, chlorinated naphthalenes; DDT, Chlordane hexachlorophene	Some by odors; eye irritant, yellow sclera of eye (liver damage)
Aliphatic hydrocarbons	Wide range: inert to anesthetic to liver/kidney damage	Variable—nonflammable to high	Variable—low to high	Largest group in use. Solvent, degreasing, fire extinguishers, refrigerants, aerosol propellants	Freons, CCl_4, trichloroethylene, methylene chloride, chloroform	Some by odor, eye irritant, "drunkenness"
Alcohols	Intoxication and narcosis, visual disturbance or blindness. Respiratory failure. Inhalation and skin absorption; irritation	Variable—low to moderate	Variable—low to high	Cellulose, plastics, rubber, explosives, adhesives, inks, pharmaceuticals, paints, dyes, leather processing	Methanol, ethanol, isopropanol, butanol, amyl alcohol, diacetone alcohol	Some by odor. "Drunkenness" breath odor. Respiratory/eye irritation

TABLE 4.3 ■ (Continued)

Category	Class Toxicity	Relative Flammability[a]	Relativity Volatility[b]	Common Industrial Uses	Examples	Signs
Esters	Anesthetic and primary respiratory irritation and pulmonary edema, eye irritation. CNS effects (some phosphate esters)	Variable—low to high	Variable—low to moderate	Plastics and resins, plasticizers, lacquer solvents, medicinals, flooring, perfumes, insecticides	Ethyl acetate, vinyl acetate, dibutylphthalate, n-butylformate, ethyliodoacetate	Some have fruit and vegetable odors which are pleasant. Eye/respiratory irritation
Ethers	Anesthetic; some are respiratory irritants; some cause liver, kidney secondary effects	Variable—some very high	Variable—some very high	Anesthetics, dewaxing solvents, refrigerants, plastics, dyes, pharmaceuticals, explosives, fuels, hydraulic fluids	Ethyl ether, methyl ether, cellulose ethers	Odor, irritation, drowsiness. Some have distinctive odor—"horseradish" (allyl ether)
Ketones	Narcotic; eye, skin, respiratory irritants. Defat skin. Some liver, kidney effects. Some skin absorption	Variable—low to high	Variable—low to high	Solvents, chemical intermediates, perfumes, explosives, plastics, dyes, adhesives, paints	Acetone, MEK, MIBK, DIBK, acetophenone, isophenone, methyl cyclohexanone	Breath odor. Some have "peppermint" fruity odors, others "ketone" odor

Paraffins	Some: simple asphyxiates Pentane-octane: narcosis, respiratory irritants	High	Some are gases at 25°C; to octane—high; above octane—low	Fuels, refrigerants, propellants	Methane, ethane, propane, butane, pentane, heptane, octane	Some are odorless; some have faint odor
Naphthenes	Anesthetics, CNS depressants, defat skin, respiratory limits	Variable—low to very high	Variable—low to high	Petroleum, liquid fuels, solvents, lubricants, chemical intermediates	Cyclohexane, turpentines, decalin	Eye/respiratory irritation. Some by odor

[a]Flash point: Low, <20°F; moderate, 20–80°F; high, >80°F.
[b]Vapor pressure: Low, <10 mm; moderate, 10–100 mm; high, >100 mm VP in mm Hg @ 25°C.

Uranium is another nephrotoxin which also damages the cells of the kidney.

Nervous System Many chemicals affect the nervous system. Often a particular chemical may be better known for producing toxicity or symptoms of acute poisoning in other organs; and yet this same chemical may also be capable of inducing chronic disorders in the nervous system. Keep in mind the following:

- The nervous system and endocrine system act together to control the functions of the rest of the body's organs.
- The nervous system is unique in its complexity and its variety of control. It consists of the peripheral nerves and the central nervous system (i.e., spinal cord brain). Within these two subdivisions are numerous cell types with specific and differing functions.
- Neurotoxins are diverse compounds that may be toxic to specific regions, specific cell types, and specific cell functions within the nervous system.

The complexity of the nervous system and the selectivity of various neurotoxins together account for a wide variety of toxic effects after chemical exposure and injury. Some are capable of fairly selective injury and may damage only hearing or sight, or specific portions of the brain or peripheral nerves.

Agents that disrupt nerve transmission are as follows:

Blocking Agents: Botulinum toxin and tetrodotoxin (produced by the puffer fish)

Depolarizing Agents: Dichlorodiphenyl trichloroethane (DDT) and pyrethrins (pesticides)

Stimulants (increase excitability): Xanthines such as caffeine, strychnine, organophosphate insecticides (parathion, malathion, diazinon, etc.), and carbamate insecticides (carbaryl or sevin, aldicarb, etc.)

Depressants (decrease excitability): Volatile organics such as halothane, methyl chloride, carbon tetrachloride, and butane.

Anoxia-causing agents and their effects include the following:

Asphyxial Anoxia (inadequate oxygen supply): Carbon dioxide, carbon monoxide, methylene chloride, and so on.

Cytotoxic Anoxia (interference with cellular metabolism): Cyanide, hydrogen sulfide, azide, dinitrophenol, malononitrile, and so on.

Agents causing physical damage to the nervous system and the specific effects include:

Those That Damage Myelin Sheath: Chronic cyanide or carbon monoxide, cyanate, lead, tellurium, thallium, triethytin, and so on.

Those That Damage Peripheral Motor Nerve: Arsenic, azide, carbon disulfide, chlorodinitrobenzene, dinitrobenzene, ethylene glycol, hexane, methanol, methyl-*N*-butyl ketone, methyl mercury, tetraethyl lead, and so on.

Other agents such as DDT and mercury cause permanent physical damage to the brain.

Blood and Hematopoietic or Blood-Forming System

Hypoxia (Anoxia) and Chemical Asphyxiants: Carbon monoxide, nitrates, nitrites, and aromatic amine compounds

Cytotoxic Hypoxia: Cyanide

Thrombocytopenia (reduction in number of platelets): Aspirin, benzene, lindane, mercurials, toluene diisocyanate, trinitrotoluene, and so on.

Agranulocytosis (reduction in white blood cells): Alkylating agents, arsenicals, aspirin, benzene, carbon tetrachloride, chlordane, lindane, trinitrotoluene, and so on.

Hemolytic Anemia and Hemolysis (decrease in red blood cells caused by lysis or destruction): Arsine, aspirin, benzene, insecticides, lead, naphthalene, nitrobenzene, penicillin, phenacetin, trinitrotoluene, and so on.

Lungs   Atmospheric pollutants including the following:

Sulfur Oxides: High concentrations can damage the alveolar septa and cause pulmonary edema. Chronic exposure to lower concentrations can cause chronic bronchitis and lead to pulmonary emphysema.

Photochemical Oxidants (produced by the action of sunlight on hydrocarbons and nitrogen oxides in air): Exposure to ozone, aldehydes, or acrolein can irritate the mucous membranes of the body and cause inflammation of the eyes and upper respiratory tract.

Nitrous Oxides: These gases can cause direct damage to the mucosal membranes of the eyes, the upper respiratory tract, or the tracheobronchial tree and alveoli. The site of damage depends on the concentration and duration of exposure.

Many chemical vapors and metal fumes have been identified as respiratory toxins.

Chemical vapors that cause direct damage to exposed respiratory system tissues arise from chlorine gas, ammonia, sulfur oxides, nitrogen oxides, and hydrogen sulfide gas. Site of injury depends on concentration and duration of exposure and the solubility of the gas in the body fluids.

Metal fumes have been identified as respiratory toxins. These include cadmium, nickel, mercury, and many others. As with chemical vapors, the site of damage within the respiratory system depends on the concentration of the metal fumes, the duration of exposure, the particle size, and the chemical characteristics of the chemical itself. The damage can occur in the upper airway, the tracheobronchial tree, or the alveolar level; it can cause either the sudden onset of symptoms or the development of symptoms over a period of months or years.

Occupationally acquired lung diseases occur through a variety of causes:

Silicosis: Exposure to quartz rock dust. This condition affects individuals who have worked in rock and slate quarries, in sandblasting, and in other industrial settings where quartz dust is present in respirable size.

Asbestosis: Exposure to asbestos fibers. This condition affects individuals who have worked in the insulating and fireproofing trades. In past years, asbestos was used widely; it is quite ubiquitous at the present time. Certain construction jobs also bring individuals into close contact with asbestos. These potentially hazardous jobs include asbestos abatement, building rehabilitation, and building demolition.

Byssinosis: An occupational lung disease seen in textile workers exposed to cotton, flax, or hemp. Symptoms include chest tightness, wheezing, and shortness of breath in some workers. In the early stages of the disease, the symptoms occur intermittently; the workers show no permanent pulmonary impairment. If individuals are removed from exposure or use effective respiratory protection, they will suffer no long-term permanent impairment. If corrective steps are not taken, however, the worker with byssinosis will develop permanent airway constriction indistinguishable from chronic bronchitic airway narrowing.

Occupational Asthma: Exposure to substances encountered in some occupations can affect immunologic pathways and lead to the development of asthma. Although the substances producing the asthma vary widely, the results are similar.

1. Bakery personnel who are exposed to flour dust can become sensitized to the flour dust itself or to the organic material in the flour dust. Termed *baker's asthma*, the condition is similar to classic allergic asthma.
2. Exposure to a number of wood dusts can produce asthma in some individuals. Red cedar, which is widely used in the construction industry, is the most commonly encountered dust-producing asthma. Other woods that can also lead to the development of asthma are mahogany, teak, and some exotic woods from Africa.
3. The chemical toluene diisocyanate (TDI) is widely used as an additive in the production of polyurethane varnish, polyurethane foam, and polyurethane paints. Some persons are quite sensitive to this chemical. When exposed to it, they become sensitized and develop asthma.

Skin The skin, an interface between the environment and the body, can be exposed to many hazardous chemicals at the workplace. The term "toxic" does not apply well in relation to the skin response; the terms "irritants," "allergens," or "infectious agent" may better describe the skin's response to chemical exposure.

Irritant contact dermatitis is the most common skin injury. It can occur in anyone, given an adequate exposure to the toxic material. Unlike allergic dermatitis, it does not require any previous exposure to the material. To develop a true allergy to something, a person must usually have a previous exposure to the chemical and then a subsequent reexposure several weeks, months, or years later. Only with the second or third exposure does a reaction occur. It takes time to develop a true allergy, and that is the difference between irritants and allergens.

With irritants an adequate exposure will elicit the inflammatory response from anyone. Irritants tend to elicit a host of responses, including: hives (wheals), which make the skin blotchy red (an erythema); purpura, a bruiselike response; blistering, eczemas, or rashes that tend to weep or ooze; erosions in which the skin seems to be scraped away; hyperkeratosis or a simple thickening of the skin (much like a callous); pustules (looking like small infections); and skin dryness and roughness. All of these responses can be considered manifestations of an irritant contact dermatitis.

The following are common causes of contact irritant dermatitis:

Water: Work during which the skin remains wet is most likely to cause irritation, because water leaches out a fair amount of natural moisturizing factors that keep the skin soft and pliable. Up to 25% of the moisturizer the skin makes on any given day can be washed out with a simple water extraction.

Soaps: If soap is used on the skin, the loss of moisturizers will rise to 35–50%. Soap tends not to be as irritating as some organic solvents and detergents. Concentration is a factor, of course.

Alkalis: These tend to be more of a problem for the skin than acids. Epoxy resin hardeners are one of the chemicals that can produce both allergic contact dermatitis and, because of their alkalinity, an irritant contact dermatitis.

Acids: These tend to be less damaging than alkalis since the normal pH of the skin tends to be in the range of 5.5, which is slightly acidic—with 7 being neutral. Since the skin tends to be slightly more acid than the rest of the body, which has a normal pH of 7.4, substances within the range of 5.5 to 7.4 are well tolerated by the skin.

Oils: A common cause of irritant contact dermatitis among industrial employees. Oils produce a host of reactions. They contain various emulsifiers, antioxidants, anticorrosion agents, and preservatives (particularly the water-soluble oils). These additives are often the culprits, rather than the oil itself.

Organic Solvents: These are also capable of producing irritant contact dermatitis, by extracting the solvent-soluble materials from the skin. Aromatic solvents are especially irritating.

Oxidants: Irritating substances such as peroxides, benzoyl peroxide, cyclohexanone, and so on, have been shown to cause contact dermatitis.

Animal Substances: These may cause irritant dermatitis, such as in poultry workers. Workers have experienced a hivelike reaction to chicken skin or other raw meats.

Pigment Disturbances: Some people develop either an increase or decrease in pigment as a result of exposure to various chemicals. Increased pigmentation or darkening of the skin will most commonly occur after contact with coal tar compounds and other petrochemical substances. This reaction may occasionally occur after handling vegetables and fruits. Decreased pigmentation will occur after exposure to common industrial chemicals such as the paratertiary butyl phenols, catechols, and hydroquinone and its derivatives.

Ulceration: This condition can be caused by trauma, thermal or chemical burns, cutaneous infection, strong acids, and a number of chemicals, including certain chromium, beryllium, nickel, and platinum salts, calcium oxide, calcium arsenate, and calcium nitrate.

Occupational Acne: This condition results from contact with petroleum and its derivitives, coal tar products, or certain halogenated aromatic hydrocarbons. The eruption can be mild, involving either (a) localized exposure perhaps to covered areas of the body or (b) severe and generalized exposure, with acne involving almost every follicular orifice (see Table 4.3 for more details). Chloracne results from exposure to chlorine compounds and halogenated chemicals. The halogenated chemicals include polyhalogenated naphthalenes, biphenyls, dibenzofurans, and dioxins.

Eyes Organic solvents, acids, alkali, and so on, are capable of irritating or burning the eye tissues. The proper eye protection should be worn in the workplace. Solvents commonly used in industry are listed in Table 4.3.

4.3 PHYSICAL HAZARDS

Fire, noise, temperature stress, ionizing, and radiation are examples of physical stresses or hazards. The employer, supervisor, or project safety officer must be aware of these hazards because of the possible immediate or cumulative effects on the health of employees at the workplace or site.

4.3.1 Fire

Fire, a chemical reaction known as *combustion*, is nonetheless a physical hazard. The fire triangle and fire tetrahedron are useful in explaining what is necessary for combustion to occur and how to extinguish combustion.

Definitions

Fire Triangle: In order for a fire to burn, it requires at least three elements: a fuel (which may be a flammable vapor), oxygen, and heat (or source of flame or ignition). These three make up the fire triangle as shown in Figure 4.3. Fuel, oxygen, and heat (spark) in the proper proportions will allow a fire to occur. Removing any one of the three will extinguish the fire. Figure 4.3 shows the fire triangle.

Fire Tetrahedron: The three components of the classic fire triangle are adequate for interpreting the mechanism by which fire can be pre-

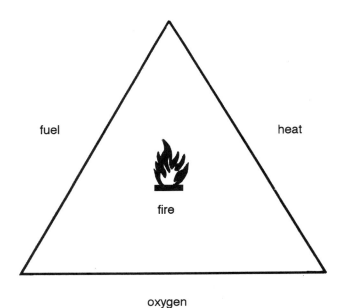

FIGURE 4.3 ■ **The fire triangle.**

vented or extinguished. However, the combustion of some materials does not depend totally on just these three factors. Hydrogen, for example, may remain mixed with chlorine in the dark for years. A chemical reaction does not occur until the mixture is exposed to sunlight. Similarly, a mixture of hydrogen and oxygen appears unreactive until it is ignited. A liquid fuel may remain in contact with oxygen for a long period of time without undergoing combustion. Futhermore, liquid fuels frequently continue to burn in an atmosphere that does not appear to contain sufficient oxygen to sustain the combustion. Thus, the fire tetrahedron adds a chemical chain reaction as another component to the fire triangle. Figure 4.4 shows how new element can be added. The addition of the chemical chain reaction allows for the conversion of the triangle to a four-sided, pyramid-shaped figure known as a *tetrahedron*. The base of the pyramid is the chemical chain reaction that caused combustion. The sides of the pyramid—reducing agent, oxidizing agent, and temperature—are the elements of the fire triangle (fuel, oxygen, and heat). The removal of one or more of the four sides will

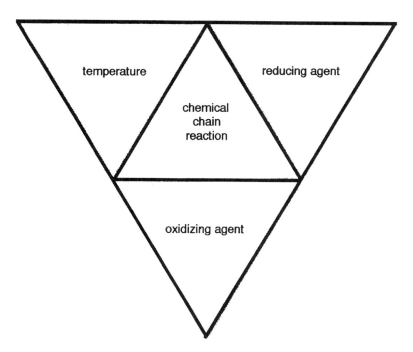

FIGURE 4.4 ■ The fire tetrahedron.

110 ■ INDUSTRIAL HYGIENE

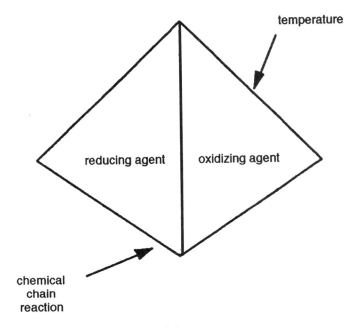

FIGURE 4.5 ■ The fire pyramid.

extinguish or prevent the fire. Figure 4.5 shows how the four necessary components form the fire pyramid.

4.3.2 Noise

Noise (see Table 4.4) is unwanted sound in the form of vibration conducted through solids, liquids, or gases; noise is also any sound that is potentially damaging to hearing or that interferes with the conduct of the worker's duties.

TABLE 4.4 ■ Typical Noise Levels

Location/Equipment	Decibel Level (dBA)
Quiet room	40
Office	70
Mechanical Equipment	80+
OSHA PEL (hearing program)	90
Rock band/jet plane	110+

The effects of noise on humans include:

- *Psychological Effects*: Noise can annoy, startle, and disrupt concentration.
- *Communication Effects*: Noise can interfere with speech and, as a consequence, affect job performance and safety.
- *Physiological Effects*: Noise can induce loss of hearing, or it can cause pain if the exposure is severe.

The risk of hearing damage does exist if the ear is subjected to high levels of noise for a sufficient period of time, some irreparable loss of hearing may occur.

A number of factors can influence the effects of noise exposure. These include the following:

- Individual variation in susceptibility
- The energy of the sound
- The frequency distribution of the sound
- Sound continuity, intermittence, or repetitive impact
- Total daily duration of exposure
- Length of employment in the noise environment

The relationship of noise and exposure time to threshold shift or hearing loss is complex. The many contributory causes compound the problem. Thus, establishing criteria for protecting workers against hearing loss is difficult. However, criteria have been developed to protect against hearing loss in the speech-frequency range. These criteria are known as the *threshold limit values for noise*.

Four nontechnical rules of thumb help to determine if the workplace has excessive noise levels:

- Is it necessary to speak very loudly or shout directly into the ear of a person in order to be understood?
- Conversation becomes difficult when the noise level exceeds 70 decibels (dBA).
- Do workers hear noises or ringing noises in their ears at the end of the workday? If so, the noise level may be excessive.
- Do employees complain that the sounds of speech or music seem muffled after leaving work, but clear in the morning when they return to work? They may be exposed to noise levels that cause a

partial, temporary loss of hearing, which can become permanent upon repeated exposures.

Permissible Levels The criteria for hearing conservation, required by the OSH Act in 29 CFR 1910.95, establish the permissible levels of harmful noise to which an employee may be subjected. The criteria specify the permissible decibel levels and hours of duration exposed to those noise levels. For example, a noise of 90 dBA is permissible for 8 hours, and 95 dBA is permissible for 4 hours.

The regulations stipulate that when employees are subjected to sound that exceeds the permissible limits, feasible administrative or engineering controls must be utilized. If such controls fail to reduce sound exposure within the permissible levels, personal protective equipment must be provided and used to reduce the sound level to permissible levels.

The regulations also require that when the sound levels exceed 85 dBA on an 8-hour, time-weighted average (TWA), a continuing, effective hearing conservation program must be administered. This program goes beyond the wearing of hearing protective equipment. Such programs require noise measurements, initiation of noise control measures, provision of hearing protection equipment, audiometric testing of employees to measure their hearing levels, and information and training programs for employees.

Hearing Loss

Temporary: Exposures to high-level noise
Short exposures to very loud noise
Recovery period not long enough

Permanent: Prolonged exposure to very loud noise
Repeated exposure to high-level noise
Aging effect

Hearing Loss Definitions

Sensorineural: Inner-ear nerve damage caused by noise exposure
Tinnitus: Ringing in the ears caused by ostosclerosis, some drugs, and very loud noise
Conductive loss: Ear blockage

Presbycusis: Nerve-cell degeneration due to aging

Temporary Threshold Shift: Measurable hearing loss after noise exposure; loss will recover typically after 16 hours

Permanent Threshold Shift: Permanent hearing loss due to noise exposure as indicated by:

- The 4000-Hz notch or decreased hearing as indicated by audiogram
- Recurrent temporary threshold shifts

Effects of Noise on Emergency Response Noise can hamper an emergency response effort because noise may interfere in communication among responders. Excessive noise may require use of hand signals or written instructions. Extra precautions may also be necessary.

4.3.3 Temperature Stress

This section covers the health effects, environmental aspects, engineering controls, protective clothing, and management of temperature stress in industry.

The extremes of temperature affect the amount and quality of work that people can do and the manner which they do it. At remediation sites, the problems can be both heat- and cold-related, depending on the season of the year and the geographic location of the work site. In industry, the problem is more often high temperatures rather than low temperatures.

Heat Stress The body continuously produces heat through its metabolic processes. Because these processes operate only within a narrow range of temperature, the body must dissipate this heat as rapidly as it is produced if it is to function efficiently.

Heat stress is a common problem; evaluating information relating the physiology of a person to the physical aspects of the environment is neither simple nor easy.

One initial question must be answered: Are the workers merely uncomfortable or are the conditions such that continued exposure will cause the body temperature to fall below or rise above safe limits?

Workers function efficiently only in a narrow body temperature range, a "core" temperature measured deep inside the body, not at the skin or at the extremities. Markedly impaired performance occurs with fluctuations in core temperatures exceeding 2°F below or 3°F above the normal core temperature of 99.6°F (37.6°C), which is 98.6°F (37.0°C) mouth temperature. If this 5° range is exceeded, a health hazard exists.

The body's regulatory processes attempt to counteract the effects of high temperature by increasing the heart rate. The capillaries in the skin also dilate, bringing more blood to the surface in order to increase the rate of cooling. Sweating and the evaporation of perspiration is an important factor in the cooling of the body.

Causes of Heat Stress. Heat stress in industry or at remediation sites has a variety of causes, including the following:

- The combination of workload and environmental heat
- Temperature and humidity factors
- Effects of chemical protective clothing and respirator use, including reduced dissipation of body heat and moisture because of protective clothing

Heat-Stress Conditions and Symptoms

Heat Rash. High temperatures can cause perspiration and softening of the outer layer of skin. This can lead to heat rash, which is common among workers exposed to hot, humid weather, electric furnaces, hot metals, or other sources of heat. The symptoms of heat rash are prickly heat, red papules, and rashes.

Heat Cramps. These can result from exposure to high temperature for a relatively long period of time; the effect is exacerbated by heavy exertion and the resulting excessive loss of salt and moisture from the body. Even if the moisture is replaced by drinking plenty of water, an excessive loss of salt can cause heat cramps or heat exhaustion.

Heat Exhaustion. This can also result from physical exertion in hot environments. Its symptoms are a mildly elevated temperature, pallor, weak pulse, headache, nausea, dizziness, weakness, profuse sweating, and cool moist skin.

Heat Stroke. This occurs by exposure to an environment in which the body is unable to cool itself sufficiently; thus, the body temperature rises rapidly. An important predisposing factor is excessive physical exertion or moderate exertion in extreme heat conditions. The methods used to control the onset of heat stroke, or the damage to the body caused by excessively high body temperature, are to reduce the temperature of the surroundings or increase the ability of the body to cool itself, thus preventing additional temperature increase. Heat stroke victims may exhibit a cessation of sweating, and the body temperature can quickly rise to fatal levels. Heat stroke is a medical emergency; the body must be cooled immediately while medical help is on the way.

Heat Stress Prevention

1. Acclimatization is the physiological adjustment of the body to working in a high-temperature environment. These progressive adjustments, occuring over periods of increasing duration, reduce the strain experienced on the initial exposure to heat. This enhanced tolerance allows a person to work effectively under conditions that might have been unendurable before acclimatization.
2. Recognize the symptoms. Being able to recognize the onset of heat stress prior to its serious consequences will assist in prevention. Watch for increased pulse rate and symptoms of heat exhaustion.
3. Drink plenty of water or a beverage that contains adequate salt.
4. Cool down and rest if symptoms occur. Figures 4.6 and 4.7 show field rest stations. These structures are set up to provide the site workers with shade, rest, and liquids to prevent heat stress.
5. Call ambulance at once, and get ice packs immediately for heat stroke.

Cold Stress Many workers encounter cold temperatures at cold-storage facilities, meat-packing plants, and remediation sites during the winter months. Because humans are warm-blooded animals (homoiotherms), they must maintain their body heat. If properly protected, they can work efficiently in both natural and man-made frigid climates as low as $-50°F$.

The human body functions best at a narrow range of body temperature (99–100°F). The body maintains this temperature by gaining heat from food and muscular activity and losing heat by radiation and sweating. The body's first physiological responses to cold are constriction of the skin's blood vessels and shivering.

116 ■ INDUSTRIAL HYGIENE

FIGURE 4.6 ■ Field rest station showing a boot and hat racks and eyewash. Before entering, workers remove and change disposable suits.

Cold temperatures first affect the skin and the peripheral blood vessels. Signals from the skin and body core are integrated in the brain (hypothalamus), which acts like a thermostat, making adjustments as needed to maintain a normal temperature.

When receiving a chill signal from the skin, the brain responds by conserving body heat and generating new heat.

Heat conservation occurs when the blood vessels in the skin or body shell constrict. This reduces heat loss from the surface of the skin and also inhibits the function of the sweat glands, preventing heat loss by evaporation.

In an effort to increase heat production, the body produces more glucose to provide additional fuel. The glucose also acts to accelerate the heartbeat, sending oxygen- and glucose-rich blood to the tissues where it is needed.

Involuntary shivering begins in an attempt to produce heat by the rapid constriction of the muscles, much as strenuous physical activity generates heat.

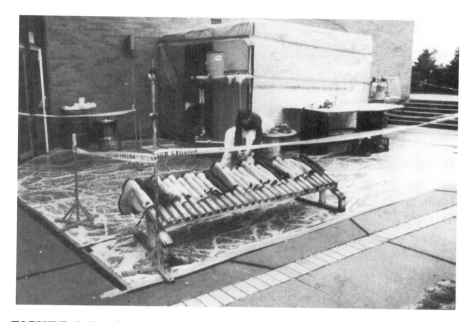

FIGURE 4.7 ■ **Rest station with boot rack in front.** *Source*: Courtesy of OHM Remediation Services Corp.

If the heat-generating and conservation activities described above are effective and the body is capable of maintaining its proper temperature, the effects of the cold will not produce serious disorders.

Cold Disorders Cold injuries are classified as either *localized*, as in frostbite, frostnip, or chillblain, or *generalized*, as in hypothermia.

The main factors contributing to cold injury are exposure to humidity and high winds, contact with wetness or metal, inadequate clothing, age, and general health.

Air temperature alone is not enough to judge the cold hazard of a particular environment. Windchill must also be taken into consideration when assessing the hazard. The influence of windchill is important and could lower significantly the effective temperature.

Hypothermia. Most cases of hypothermia develop in air temperatures between 30°F and 50°F (2–10°C). However, factors such as windchill and humidity can make the effective temperature considerably less.

The first symptoms of hypothermia are uncontrollable shivering and the sensation of cold. The heartbeat slows and sometimes becomes irregular. The pulse weakens and the blood pressure changes.

Severe shaking or muscular rigidity follow, often accompanied by slurred speech, memory lapses, incoherence, and drowsiness. Other symptoms that can occur before complete collapse are cool skin, slow, irregular breathing, low blood pressure, apparent exhaustion, and fatigue even after rest.

As the core temperature drops, the victim can become listless and confused and may make little or no effort to keep warm. Pain in the extremities can be the first warning of dangerous exposure to cold, and severe shivering must be taken as a sign of danger.

As the core temperature approaches 85°F, serious problems can develop because of significant drops in blood pressure, pulse rate, and respiration. In some cases the victim may die.

Alcohol and sedatives increase the risk of hypothermia. Alcohol dilates the blood vessels near the skin surface, which increases heat loss and lowers the body temperature. Sedatives interfere with nerve impulse transmission to the brain, also interfering with body temperature regulation.

Frostbite. Frostbite can occur without hypothermia when the extremities do not receive sufficient heat from the body core. This can result because of inadequate circulation or inadequate insulation. Frostbite occurs when fluids freeze around the cells of the body tissues. The condition results in damage to and loss of tissue. The most vulnerable parts of the body are the nose, cheeks, ears, fingers, and toes.

Damage from frostbite can be serious; scarring, tissue death, and amputation are possible.

Symptoms of frostbite include the following:

1. The skin changes color to white or grayish yellow, progresses to reddish violet, and finally turns black as the tissue dies.
2. Pain occurs first, but then subsides.
3. Blisters may appear.
4. The affected part is cold and numb.

Chilblains. This condition results from long, continuous exposure to cold without freezing, combined with persistent dampness or actual immersion in water. Edema (swelling), tingling, itching, and severe pain occur; blistering, death of skin tissue, and ulceration may follow.

Frostnip. This occurs when the face or extremities are exposed to a cold wind, causing the skin to turn white.

Preventing Cold Stress. The health and safety professional must consider factors relating to both the individual and the environment when attempting to prevent cold stress. Acclimatization, water and salt replacement, medical screening, continued medical supervision, proper work clothes, and training and education will contribute to the prevention of cold stress and injury related to work in cold environments.

Personal protective equipment and protective clothing are essential for preventing cold stress. Preserving the air space between the body and the outer layer of clothing is important in order to retain body heat. The more air pockets each layer of clothing has, the better the insulation. However, the insulation effect is negated if the clothing interferes with the evaporation of sweat or if the skin or clothing is wet.

The most important parts of the body to protect are the feet, hands, and face. Keeping the head covered is important, because as much as 40% of body heat can be lost when the head is exposed.

Clothing made of thin cotton fabric is recommended; it helps evaporate sweat by absorbing the moisture and allowing the moisture to move to the surface of the garment. Loosely fitting clothing also aids sweat evaporation. Tightly fitting clothing made of synthetic fabrics interferes with evaporation.

Recommended clothing may include a cotton t-shirt and shorts or underpants under cotton or wool thermal underwear. Socks with a high wool content are best. When two pairs of socks are worn, the inside pair should be smaller and made of cotton. Wool or thermal trousers are preferable. Trousers should be lapped over boot tops to keep out snow or water. For heavy work, a felt-lined, rubber-bottomed, leather-topped boot with a removable felt insole provides good protection. Boots should be waterproofed, and socks should be changed when they become sweat-soaked.

A wool shirt or wool sweater over a cotton shirt is desirable. Shirts and sweaters should be worn in layers. Outer coats should be loose-fitting, with a drawstring at the waist. Sleeves should fit snugly. The use of a hood is recommended, because it prevents the escape of warm air from around the neck. A wool knit cap also provides good head protection. When a hard hat is worn, a liner should be used. Wool mittens are more efficient insulators than gloves; they can be worn over gloves for extra warmth.

A face mask or scarf is vital when working in a cold wind. Workers should frequently remove face protectors, so that they can check for signs of frostbite.

Workers should wear several layers of clothing instead of a single heavy outer garment. In addition to offering better insulation, layers of clothing can be removed as needed to keep the worker from overheating. The outer layer should be windproof and waterproof. Body heat is lost quickly if the protective layer is not windproof.

All clothing and equipment must be properly fitted and worn to avoid interfering with the circulation. Full-facepiece respirators should have nose-cups to reduce fogging and frosting of the facepiece.

Since liquids conduct heat and cold better than air, a spill of cold gasoline on the skin can freeze the tissues quickly. For this reason, chemical-resistant gloves are recommended for chemical-handling operations.

The rules of survival are the same, no matter what the situation is: Think clearly, be properly prepared, and be aware of the weather forecast for your workplace or site.

4.3.4 Radiation

This section will briefly describe what radiation is, and why it poses a hazard to those working in areas where radioactive materials are stored, buried, or handled.

The human body is made up of various chemical compounds that are in turn composed of molecules and atoms. Each atom has a nucleus with its shell(s) of electrons.

When ionizing radiation interacts with body tissue, ionization of that tissue may occur, forcibly ejecting from their orbits some of the electrons surrounding the atoms. The greater the intensity of the ionizing radiation, the more ions will be created and the more physical damage will result to the cells.

Health and safety professionals should consider at least three basic factors in any approach to radiation safety:

- Radioactive materials emit energy that can damage living tissue.
- Different kinds of radioactivity present different kinds of radiation safety problems. Alpha particles, beta particles, x-ray, gamma radiation, and neutrons are among the different types of ionizing radiation that can cause problems.

- Radioactive materials can be hazardous in two different ways. Certain materials are hazardous even when located some distance away from the body. These are external hazards. Other types are hazardous only when they enter the body. Entry can occur by breathing and eating or through broken skin. These are referred to as *internal radiation hazards*.

Kinds of Radioactivity The five kinds of radioactivity that are of concern are alpha, beta, x-ray, gamma, and neutron. The first four are the most important because neutron sources are not used in ordinary manufacturing operations.

Alpha particles are identical to a helium nucleus consisting of two neutrons and two protons. These particles result from the radioactive decay of heavy elements such as uranium, plutonium, radium, and thorium. Alpha particles have a large mass as compared to most other types of particulate radiations, such as neutrons and beta particles. Because of their double positive charge, they have great ionizing power, but their large size results in very little penetrating power. Alpha particles are the least penetrating. They will not penetrate thin barriers such as paper, cellophane, and skin, but they are still of concern.

Beta particles are electrons resulting from the conversion of a neutron to a proton in the nucleus of an atom. Beta particles have a greater range and considerably more penetration power than alpha particles. An aluminum plate 1 mm thick can stop most beta particles.

The electromagnetic ionizing radiations, gamma and x-ray, have both electromagnetic wave and particle characteristics.

X-rays originate outside the nucleus of atoms. In x-ray machines, they are produced by applying a high positive voltage between the source of electrons and a collecting terminal within a vacuum tube. When the electrons strike a suitable target, such as tungsten, their energy is partly converted into x-ray energy.

Gamma rays originate from unstable atomic nuclei releasing energy to gain stability. They have definite energies, characteristic of the nuclide (element) from which they were emitted. For example, cobalt-60 gives off two gamma rays of energy around 1.25 MeV. Gamma and x-rays ionize materials indirectly through mechanisms that involve ejection of high-speed electrons from atoms that have absorbed the energy. X-rays and gamma rays are highly penetrating. The shielding required to reduce, by a factor of 10, the radiation intensity of a gamma ray emitted from cobalt-60 is 20.8 cm concrete, 6.9 cm steel, or 4.0 cm lead. The shielding requirements vary with the energy of the gamma emitted and/or the source.

Both x-rays and gamma rays can pose severe hazards. X-rays and gamma rays can penetrate thick layers of lead and concrete, so it is difficult to shield the body against them. They can also penetrate great distances through air and may therefore constitute a hazard even far from a source of radiation. Finally, these electromagnetic waves can easily penetrate the skin and irradiate organs within the body; in fact, they can irradiate the whole body. However, since they have virtually no mass, they are less of an internal hazard than either alpha or beta radiation.

Neutrons are very penetrating. Because they have no electrical charge, they do not interact with electrons. They do interact with nuclei of atoms, producing charged particles that can ionize or excite other atoms. Neutrons have a short half-life of 10.6 minutes, and they decay into a proton and an electron. Even though their half-life is short, neutrons are relatively large particles and can cause severe damage to living tissues. They therefore pose a great hazard for workers who have a potential exposure during the performance of their duties.

External Versus Internal Hazards Radioactive materials that emit x-rays, gamma rays, or neutrons are external hazards. In other words, such materials can be located some distance from the body and emit radiation that will produce ionization and thus damage as it passes through the body. To provide protection against external radiation, three safety practices can be employed:

- Limit exposure time. Dose is directly proportional to the time spent in the field.
- Work at a safe distance. The radiation dose received is inversely proportional to the square of the distance of separation.
- Use barriers or shielding.

The use of a combination of these practices is essential for adequate protection against external radiation hazards.

As long as a radioactive material that emits alpha particles remains outside the body, it will not cause trouble. Internally, however, such material is a hazard because of the ionizing ability of alpha particles when in close proximately to soft tissue. Once inside the body—in the lungs, in the stomach, or as an open wound, for example—there is no thick layer of skin to serve as a barrier. Thus damage results. Alpha-emitting materials that will accumulate as persisting deposits in specific parts of the body (strontium-90 deposited in bone) are very hazardous.

Beta-emitters are considered to be an internal hazard. However, they can also be classified as an external hazard because they can produce burns when in contact with skin.

REVIEW QUESTIONS

1. How does the solubility of an irritant gas, vapor, or particulate affect its ability to penetrate deep into the respiratory tract?
2. What are the most common ways that substances enter the body? List five chemicals that enter the body by each route.
3. Acids and alkali are hazardous because they cause acute effects. What are the target organs of these affects?
4. What is a TVL-TWA? Using your TLV book, look up the TLVs for the following:
    1. Benzene   _____
    2. Toluene   _____
    3. Xylene   _____
    4. Diazinon   _____
    5. 2-Butanone   _____
    6. Sulfuric acid   _____
    7. Trichloroethylene _____
    8. Tetraethyl lead _____
    9. Arsenic   _____
    10. Chlordane   _____

Why are some TLVs in ppm and others in mg/m^3?

What does the skin notation mean next to the substance name in the TLV book?

What do the following symbols mean?

++

Δ

5. Describe the importance of the following terms:
   - Boiling point
   - Flammability
   - Flash point
   - Vapor density
6. What is the regulatory significance of the following decibel levels?
   - 85 dBA
   - 90 dBA
7. What is the prime hazard of radioactive materials that are alpha-emitters?

REFERENCES

American Conference of Governmental Industrial Hygienists, *Threshold Limit Values for Chemical Substances and Physical Agents and Biological Exposure Indices with Intended Changes for 1994–1995*, ACGIH, Cincinnati, OH, 1995.

Klaassen, C. D., M. O. Amdur, and J. Doull, eds., *Casrett and Doull's Toxicology—The Basic Science of Poisons*, 3rd ed., Macmillan, New York, 1986.

Plog, B. A., ed., *Fundamentals of Industrial Hygiene*, 3rd ed., National Safety Council, Chicago, IL, 1988.

Meyer, E., *Chemistry of Hazardous Materials*, 2nd ed., Prentice-Hall, Englewood Cliffs, NJ, 1987.

U.S. Department of Health and Human Services, *NIOSH Pocket Guide to Chemical Hazards*, U.S. Government Printing Office, Washington, D.C., 1994.

U.S. Department of Transportation, *1990 Emergency Response Guidebook*, Department of Transportation, DOT P 5800.5, Washington, D.C., 1990.

U.S. Environmental Protection Agency, *EPA Toxicology Handbook*, 2nd ed., Government Institutes, Rockville, MD, 1987.

HAZARD RECOGNITION

This chapter will focus on hazard recognition in both controlled and uncontrolled situations. This discussion will include routine methods, such as placards, labels, and other identifiers, used to recognize hazardous materials. Methods for gaining more information about specific chemicals, such as from material safety data sheets (MSDS), the *NIOSH Pocket Guide to Chemical Hazards*, and other resources, will also receive attention.

CHAPTER OBJECTIVES

When you have completed this chapter, you will be better able to

- Recognize the major types of hazards
- Recognize hazardous materials placards, labels, and container shapes
- Review the use of the Department of Transportation (DOT) handbook to identify different types of hazardous materials
- Describe the many different types of drums and their purposes and how to handle these drums properly
- Recognize the types of information presented by shipping papers and MSDS

5.1 INTRODUCTION

In an industrial setting, most materials used at a plant are known ahead of time. However, materials may be accidentally mislabeled or delivered to the wrong site, so unknown or unexpected hazards may also exist in the setting. By understanding each of the sources of information discussed in this section, employees may be prepared to identify these potential hazards.

In an uncontrolled setting such as a Superfund site, abandoned landfill or disposal site, remediation site, or other sites where activities can expose workers to unknown chemicals or mixtures of chemicals, the ability to identify or locate other sources of information is very important. Such information not only helps workers to identify the potential hazards of that site, it also helps workers prepare for those hazards and reduce the risk of injury.

5.2 RECOGNIZING THE TYPES OF HAZARDS

Information in this section concerns the basic hazards that may exist and general methods of determining the nature of the hazard. Specific hazards at a plant site are described in the plant's emergency response plan, and known hazards associated with a Superfund or remediation site are described in the site safety plan. Specific hazards not included in this section may require additional training.

Health and safety hazards can be grouped into three main types:

- Chemical
- Biological
- Physical

Examples of each type of hazard are listed below:

Chemical
- Flammable liquids and vapors
- Toxic gases and liquids
- Carcinogens (cancer-causing)
- Corrosives
- Poisons

Biological
- Infectious hospital wastes
- Research materials

5.2 RECOGNIZING THE TYPES OF HAZARDS ■ 127

Physical   Slips, trips, and falls
                   Noise
                   Electricity
                   Temperature stress
                   Radiation

5.2.1 DOT—Labels and Placards

Part of preplanning involves understanding the systems that are available to identify hazardous materials. Identification information is at hand on labels affixed to small containers (drums, packages, boxes) and on placards affixed to large containers (trailers, rail cars, tanks). Several different systems exist; one or more may be used at the plant or site where the materials were stored by personnel or companies that handled or supplied the materials. Some of these systems are described below.

The Department of Transportation (DOT) System of Placarding and Labeling What does the DOT system look like?

- It may be *diamond-shaped.*
- It may be *color-coded*:

Color	*Hazard*
Orange	Explosive
Red	Flammable
Green	Nonflammable
Yellow	Reactive
White	Poisonous
White/red vertical stripes	Flammable solid
White top with black bottom	Corrosive
Two colors	Two major hazards

- It may be *word-coded (hazard class name)*:

    Explosives

    Blasting agents

    Dangerous (may be used with *mixed loads*)

- It may be symbol-coded

Symbol	Hazard
Bursting ball	Explosive
Flame	Flammable
W with slash	Dangerous when wet
Skull and crossbones	Poisonous
Circle with flame	Oxidizing material
Cylinder	Nonflammable gas
Propeller	Radioactive
Test tube/hand/metal	Corrosive
Special symbol	Infectious (discussed later)

- It may be *number-coded*. A four-digit number in the center identifies a specific compound. These numbers (UN/UA) are identified in the DOT *Emergency Response Guidebook*. For example, 1203 is gasoline. A one-digit number at the bottom is the UN (United Nations) hazard class. The DOT hazard class designations are listed in Table 5.1. The placard that follows is red with white symbols except for the black four-digit number. This placard shows that the substance is flammable (the flame and the red background), a flammable liquid (the UN class number 3), and acetone with the four-digit number 1090 as shown in Figure 5.1. The acetone placard is called a "number placard," which means that the number at the center of the placard

TABLE 5.1 ■ UN Hazard Class Designations

1	Explosives
2	Gases (compressed, liquified, or dissolved under pressure)
3	Flammable liquids
4	Flammable solids or materials
5	Oxidizing materials
6	Poison and infectious materials
7	Radioactive materials
8	Corrosives
9	Miscellaneous dangerous materials and other regulated materials (ORM-D)

5.2 RECOGNIZING THE TYPES OF HAZARDS ■ 129

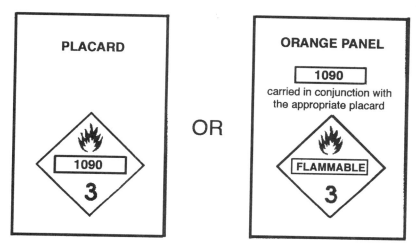

FIGURE 5.1 ■ The proper placarding for acetone.

specifies the exact contents of the container. "Word placards" are so named because a word designating a type of hazard (e.g., "flammable") will be printed in the middle of the placard. Number placards must be displayed on large portable tanks, tank trucks, and rail cars. A word placard means that drums or smaller containers are present.

You can find more information on these numbers and symbols by consulting the *Emergency Response Guidebook*, Federal DOT publication number P 5800.5.

To use the DOT book, you need to know either the chemical name, the identification number, or the appearance of the placard.

- *Yellow Section—Listing of Chemicals by ID Number.* If you know the *number*, look in the yellow pages to find the name guide number.
- *Blue Section—Listing of Chemicals by Name.* If you know the *name*, look in the blue pages to find the ID number and guide number.
- *Orange Section—Guides (11–76).* Once you have obtained the guide number from either the yellow or the blue sections, you can proceed to the designated guide that contains the appropriate hazard warnings, emergency response, and first aid information.

- *Green Section—Listing of Those Chemicals Requiring Isolation and Evacuation.* This section lists chemicals and the isolation and evacuation distances recommended in an emergency response to small and large spills.

The Table of Placards This table is useful when the shipping papers, manifest or driver are not available to assist in the identification of the materials involved in the incident. This table also serves as a reference when responders must compare represented placards with the placard on the vehicle, thus helping to identify the hazards associated with the load.

Summary of DOT Guidebook Use Start at page i.

1a. Find the placard ID number (4-digit), or
1b. Find the chemical name of the material.
2. Look up material's 2-digit guide number (11–76):
- Yellow section by ID number
- Blue section by chemical name

3. Turn to appropriate guide
- Orange section (guide 11 through 76)

4. Follow instructions.

Using the guidebook and the number placard in Figure 5.2, determine the contents of the vehicle carrying this placard.

What is the truck in Figure 5.3 carrying? What is the truck in Figure 5.4 carrying?

5.2.2 The NFPA-704M System

The National Fire Protection Association (NFPA) system is used on storage vessels and containers. What does the NFPA system look like?

- It is *diamond-shaped.*
- It is *color-coded in four small diamonds:*

Color	Hazard
Red	Flammability
Blue	Health
Yellow	Reactivity
White	Special information (such as radioactive)

5.2 RECOGNIZING THE TYPES OF HAZARDS ■ **131**

FIGURE 5.2 ■ The placard for paint, enamel, lacquer, polish, or shellac.

What is this truck carrying?

FIGURE 5.3 ■ Truck displaying the placard for gasoline.

132 ■ HAZARD RECOGNITION

FIGURE 5.4 ■ Truck displaying the placard for sodium hydroxide.

- It is *number-coded in the red, blue, and yellow diamonds*. This type of coding ranks the potential health, flammability, and reactivity hazard (see Figure 5.5). The ranges are from 0 (least hazardous) to 4 (most hazardous). The special information (white) section of the NFPA-704M label may contain symbols that give more information about the chemical. Figure 5.6 shows examples of some of these symbols and their meanings.

5.2.3 The HMIS (Hazardous Material Information System)

These labels are used on storage vessels and containers (see Figure 5.7). What does the HMIS system look like?

- It is *rectangular*.
- It is *color-coded:*

  Color   Hazard
  Blue   Health risk
  Red   Flammability
  Yellow   Reactivity
  White   Personal protective
           equipment needed

- It is *number-coded*. This coding ranks the potential health, flammability, and reactivity hazard. The ranges from 0 (least hazardous) to 4 (most hazardous).

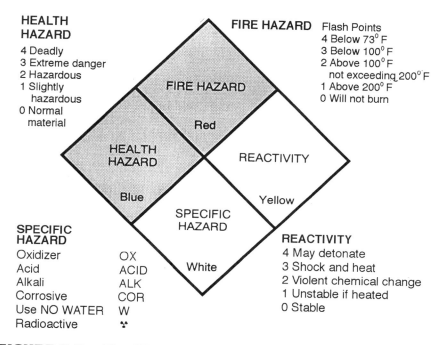

FIGURE 5.5 ■ **The NFPA-704M diamond.** *Source:* Copyright © 1990, National Fire Protection Association, Quincy, MA 02269. This warning system is intended to be interpreted and applied only by properly trained individuals to identify fire, health and reactivity hazards of chemicals. The user is referred to certain limited number of chemicals with recommended classifications in NFPA 49 and NFPA 325M which would be used as a guideline only. Whether the chemicals are classified by NFPA or not, anyone using the 704 system to classify chemicals does so at their own risk.

| Do not use water | Biological Hazard | Oxidizer | Radiation Hazard |

FIGURE 5.6 ■ The special information (white) section of the NFPA-704M label may contain additional information about the hazards of the chemical.

134 ■ HAZARD RECOGNITION

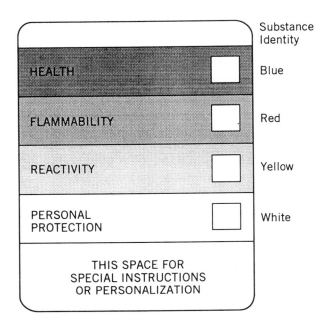

FIGURE 5.7 ■ The Hazardous Material Information System (HMIS).

- It is *letter-coded*. This coding tells what personal protective equipment to use when working with the material. The ranges from A (safety glasses) to K (full protective suit with gloves, boots, a hood or mask, and an air line or self-contained breathing apparatus). If the personal protection is coded X, specialized handling procedures are needed.

5.2.4 Infectious Materials

The most common type of packaged biological waste is probably infectious waste from a hospital or other health care facility. This type of waste should be in boxes, plastic containers, or red plastic bags. These containers should be marked on all sides with the fluorescent orange infectious materials symbol shown in Figure 5.8. Examples of infectious materials include used needles and syringes, soiled bandages, test tubes, and disposable vials.

As with any hazardous waste, disposal containers may break open in transportation accidents. If you notice anything that looks like hospital

INFECTIOUS MATERIALS

FIGURE 5.8 ■ Infectious material label.

waste lying around an accident site, move away from the area. Do not pick up or touch the material.

5.2.5 Recognizing Chemical Hazards
Drums
Container Labels, Materials of Construction, and Configuration
For many years, hazardous and nonhazardous wastes have been stored, transported, and disposed of in drums. Many remediation sites contain large quantities of buried, abandoned and corroding drums. In most instances the drums are in poor condition and often lack labels or identifying information. As with most metal or steel containers, over time most drums do corrode, especially those buried and exposed to moisture. Thus, workers should excercise care prior to handling any drum.

At remediation sites containing large quantities of drums, most of the activities focus on the sampling, identification, and bulking of drum contents for transport and subsequent disposal.

At facilities that generate hazardous waste, the identity of the contents of the containers may be generally known; however, in an emergency the labels or placards may be damaged or blocked from view. Therefore, it can prove important to be able to recognize specific types of containers.

136 ■ HAZARD RECOGNITION

Contents of Drums The "clues" to contents of drums come from the material from which the drums are made as well as whether the drum is closed-top or open-top. Closed-top drums are sealed drums that have small openings in the top of the drum (e.g., the bung opening), through which liquids can be poured. Open-top drums have removable lids and do not have the small openings characteristic of the closed-top drum. Open-top drums are used to package solid materials (e.g., contaminated soil, garments, tools, and other contaminated materials) for transportation to the proper disposal facilities. Figure 5.9 shows the two drum types. The open-top drum is in the foreground, and the close-top drum with the bung holes is in the background.

Some types of drums and what they contain are listed below:

1. Closed-top metal drums with a bung hole normally contain noncorrosive products in liquid form.
2. Closed-top plastic or composite (plastic liner inside metal or cardboard) drums usually contain corrosive liquids.

FIGURE 5.9 ■ The two drum types. The open-top drum is in the foreground, and the close-top drum with the bung holes is in the background.

3. Open-top metal drums usually contain noncorrosive solids or sludges.
4. Open-top plastic drums usually contain corrosive solids or sludges.
5. Other types of drums such as stainless steel, nickel, and MONEL™ are used for chemicals that require special containers for safe containment. These containers usually can be recognized by their metallic color.

Handling and Inspecting Drums and Other Containers   The appropriate procedures for handling drums depend on the drum contents. Thus, prior to any handling, workers should visually inspect the drums and attempt to identify the contents or the general hazard(s) posed by the drum contents. As stated previously, many drums are in poor condition and may fail during the initial phases of drum sampling or handling. Figures 5.10 and 5.11 show several examples of drums and other containers in poor condition. These drums will be overpacked in salvage drums (85 gallon) which are larger and made of the same material as the original drum (see Figure 10.4).

The following list contains important features to look for during the inspection. These will assist you in recognizing the potential hazards involved in the subsequent drum-related activities.

FIGURE 5.10 ■ Corroded drums.

138 ■ HAZARD RECOGNITION

FIGURE 5.11 ■ Poorly labeled drums, in poor condition, stacked without a pallet.

- Symbols, words, or other marks on the drum indicating that its contents are hazardous, (e.g., radioactive, explosive, corrosive, toxic, flammable)
- Symbols, words, or other marks indicating that the drum contains discarded laboratory chemicals, reagents, or other potentially dangerous materials in small-volume individual containers
- Signs of deterioration such as corrosion, rust, and leaks
- Signs that the drum is under pressure (e.g., swelling and bulging)
- Drum type (see Table 5.2)
- Configuration of the drum head (removable lid, closed, or lined)

Special Drums and Their Associated Hazards Polyethylene or polyvinyl chloride (PVC)-lined drums often contain strong acids or bases. If the lining is punctured, the substance usually quickly corrodes the steel, resulting in a significant leak or spill.

Exotic metal drums (e.g., aluminum, nickel, stainless steel, or other unusual metal) are very expensive and usually contain an extremely dangerous material.

TABLE 5.2 ■ Drum Characterization

Observation	Probable Contents	Hazard	Personal Protective Equipment (PPE)
Corroded container particularly around bung or filler holes or in a running pattern down the sides	Corrosive	Corrosive to skin and eyes	Complete skin protection
Plastic-type container		Incompatible with alkaline solutions	Acid cartridge, air purifying respirator
Plastic-lined drum (poly-lined)		Releases toxic or explosive gases when combined with halogenated chemicals	Goggles or face shield
Black and white hazardous warning label			Rubber boots
Stinging, burning sensation on skin or in eyes, nose, throat			Acid-resistant gloves (butyl rubber)
Turn litmus paper red or blue			
Has a pH above 12.5 or below 2.0			
Causes soda ash (sodium bicarbonate) to bubble vigorously.			
Fiber container	Oxidizer	Highly reactive when mixed with organics	High-efficiency particulate respirator

TABLE 5.2 ■ Continued

Observation	Probable Contents	Hazard	Personal Protective Equipment (PPE)
Small drum with poly-liner			
Yellow warning label			Goggles or face shield
Stainless steel container	Explosive, reactive or flammable	Extreme hazard precautions	Extreme hazard precautions
Bulging top and/or sides	Flammable liquids	Fire, explosion; keep from oxidizers	Nonsparking tools
Vapor-pressure release when opened		Toxic fumes	Organic vapor respirator or SCBA
Strong, generally pleasant odor			Protective clothing
Float or water			Monitoring equipment
Evaporates quickly			Organic vapor analyzer, combustible gas indicator, oxygen meter
Pressure release bung			
Red warning label			Polyvinyl chloride (PVC) splash gear

Characteristics	Substance	Hazard	Equipment
Little or no vapor pressure	Halogenated solvents	Toxic, combustible	Organic vapor respirator
Sweet-smelling			Goggles or face shield
Oil or grease sludge at bottom			Protective clothing
No signs of corrosion around filler holes			PVC boots
No poly-liner			Oxygen meter
Floats on water			Organic analyzer
Little or no vapor pressure	Heavy metals	Toxic	Particulate respirator
No strong odor	Groundwater	Contamination	Protective clothing
Sludge at bottom (often varied colors)			Splash goggles
Slow evaporation			Boots
Mixes with water			Gloves

Singled-walled drums are often used as a pressure vessel. These drums have fittings for both product filling and placement of an inert gas, such as nitrogen. May contain reactive, flammable, or explosive substances.

Laboratory packs are used for disposal of expired chemicals and process samples from university laboratories, hospitals, and similar institutions. Individual containers within the lab pack are often packed in absorbent material. They may contain incompatible materials, radioisotopes, or shock-sensitive, highly volatile, highly corrosive, or very toxic exotic chemicals. Laboratory packs can be an ignition source for fires at hazardous waste sites.

Drum Handling Safety The following procedures and equipment will maximize worker safety during drum handling and movement:

- Train personnel in proper lifting and moving techniques to prevent back injuries.
- Make sure the vehicle selected to convey drums has sufficiently rated load capacity to handle the anticipated loads, and make sure the vehicle can operate smoothly on the available road surface.
- Air-condition the cabs of vehicles to increase operator efficiency; protect the operator with heavy splash shields.
- Supply operators with appropriate respiratory protective equipment when needed. Normally, a combination SCBA/SAR with an air tank fastened to the vehicle—or, alternatively, an airline respirator and an escape or egress bottle—is used because of the high potential hazards of drum handling. This improves operator efficiency and provides protection in case the operator must abandon the equipment.
- Have overpack or salvage drums ready before any attempt is made to move drums.
- Before moving anything, determine the most appropriate sequence in which the various drums and other containers should be moved. For example, small containers may have to be moved first to permit heavy equipment to enter and move the drums.
- Exercise extreme caution in handling drums that are not intact and tightly sealed.
- Ensure that operators have a clear view of the roadway when carrying drums. Where necessary, have ground workers available to guide the operator's motion.

5.2 RECOGNIZING THE TYPES OF HAZARDS ■ 143

Drums Containing Packaged Laboratory Wastes (Lab Packs) Laboratory packs (i.e., drums containing individual containers of laboratory materials normally surrounded by cushioning absorbent material as shown in Figure 5.12) can be a dangerous ignition source for fires at hazardous waste sites. They sometimes contain shock-sensitive materials. Such containers should be considered to hold explosive or shock-sensitive wastes until they are characterized. The classification system listed in Table 5.3 should be followed. If handling is required, the following precautions are among those that should be taken:

- Prior to handling or transporting lab packs, make sure all nonessential personnel have moved a safe distance away.

FIGURE 5.12 ■ **Laboratory pack showing hazardous waste label and hazard class label.**

TABLE 5.3 ■ Example of Lab Pack Content Classification System for Disposal

Classification	Example
Inorganic acids	Hydrochloric, sulfuric
Inorganic bases	Sodium hydroxide, potassium hydroxide
Strong oxidizing agents	Ammonium nitrate, barium nitrate, sodium chlorate, sodium peroxide
Strong reducing agents	Sodium thiosulfate, oxalic acid, sodium sulfide
Anhydrous organics and organometallics	Tetraethyl lead, phenylmercuric chloride
Anhydrous inorganics and hydrides	Potassium hydride, sodium metal hydride, sodium and potassium metal
Toxic organics	PCBs, insecticides
Flammable organics	Acetone, hexane, toluene
Inorganics	Sodium carbonate, potassium chloride
Inorganic cyanides	Potassium, sodium and copper cyanide
Organic cyanides	Cyanoacetamine
Toxic metals	Arsenic, cadmium, lead, mercury

- Whenever possible, use a grappler or specific drum handling equipment constructed for explosive containment for initial handling of such drums.
- Maintain continuous communication with the site safety officer and/or the command post until handling operations are completed.
- Once the lab pack has been opened, have a chemist inspect, classify, and segregate the bottles within it, without opening them, according to the hazards of the wastes. An example of a system for classifying lab pack wastes is provided in Table 5.3. The objective of classification is to segregate the contents into groups of compatible chemicals that can be repacked for transit and/or disposal. Be sure to pack the bottles and other containers with sufficient cushioning and absorption materials to prevent excessive movement of the containers and to absorb all liquid in case of a spill or leak. Ship the repack to an approved disposal facility.

- If crystalline material is noted at the neck of any bottle, handle it as a shock-sensitive waste, and get expert advice before attempting to handle it. Special care should be exercised due to the potential presence of picric acid or other similar material.
- Palletize the repacked drums prior to transport and make sure the drums are secured to the pallets.

Bulging Drums The following procedures should be followed when inspecting and handling bulging drums:

- Pressurized drums are extremely hazardous. Wherever possible, do not move drums that may be under internal pressure, as evidenced by bulging or swelling.
- If a pressurized drum has to be moved, whenever possible, handle the drum with a grappler unit constructed for explosive containment. Move the bulged drum only as far as necessary to allow seating on firm ground, or carefully overpack the drum. Exercise extreme caution when working with or adjacent to potentially pressurized drums.

Handling Leaking, Open, and Deteriorated Drums If a drum containing a liquid cannot be moved without rupture, immediately transfer its contents to a sound drum using a pump designed for transferring that liquid.

Using a drum grappler, immediately place the emptied drum in an overpack container.

Procedures for Opening Drums Containing Hazardous Wastes

- If a supplied-air respiratory protection system is used, place a bank of air cylinders outside the work area (when possible) and supply air to operators via airlines. This type of respiratory protection equipment enables the workers to operate in relative comfort for extended periods of time.
- Protect personnel by keeping them a safe distance from drums being opened. If personnel must be located near the drums, place explosion-resistant plastic shields between them and the drums. Controls for drum opening equipment, monitoring equipment, and fire suppression equipment should be located behind explosion-resistant shielding.

- If possible, monitor continuously during opening. Place sensors or monitoring equipment such as colorimetric tubes, dosimeters, radiation survey instruments, combustible gas meters, organic vapor analyzers, and oxygen meters as close as possible to the source of contaminants (i.e., at the drum opening).
- Use the following remote-controlled devices for opening drums: Drum handling

   Pneumatically operated impact wrench to remove drum bungs

   Hydraulically or pneumatically operated pierces (see Figure 5.13)

   Backhoes equipped with bronze spikes for penetrating drum tops in large-scale operations (see Figure 5.14)

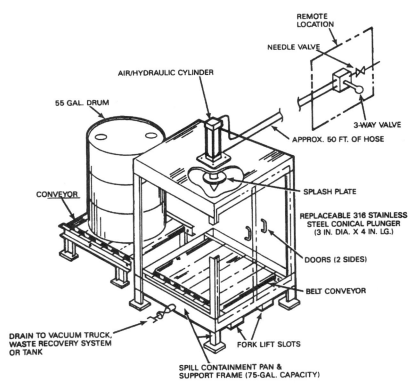

FIGURE 5.13 ■ **Hydraulic-operating single-drum puncture device.** *Source:* (NIOSH, *Occupational Safety and Health Guidance Manual for Hazardous Waste Site Activities*, 1985.

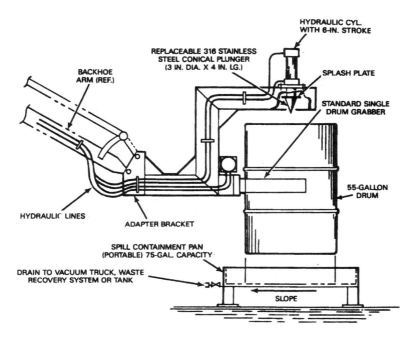

FIGURE 5.14 ■ **Backhoe mounted-drum puncture device.** *Source:* NIOSH, Occupational Safety and Health Guidance Manual for Hazardous Waste Site Activities, 1985.

- Do not use picks, chisels, and firearms to open drums.
- Hang or balance the drum-opening equipment to minimize worker exertion.
- If the drums show signs of swelling or bulging, perform all steps slowly. Relieve excess pressure prior to opening; if possible, perform this procedure from a remote location using such devices as a pneumatic impact wrench or hydraulic penetration device. If pressure must be relieved manually, place between the worker and the drum's bung a barrier such as explosion-resistant plastic sheeting to deflect any gas, liquid, or solids that may be expelled as the bung is loosened.
- Remove or drill through the bung to open exotic metal drums and polyethylene or PVC-lined drums. Exercise extreme caution when manipulating these containers.

- Do not open or sample individual containers within laboratory packs.
- Reseal open bungs and drill openings as soon as possible, using new bungs or plugs to avoid explosions and/or vapor generation. If open drums cannot be resealed, place the drum in an overpack. Plug any openings in pressurized drums, using pressure-venting caps set to 5-psi (pounds per square inch) release to allow venting of vapor pressure.
- Decontaminate equipment after each use to avoid mixing incompatible wastes.

Drum Sampling Drum sampling can be one of the most hazardous activities to worker safety and health because this procedure often involves direct contact with unidentified wastes. Prior to collecting any sample, develop a sampling plan:

- Research background information about waste.
- Determine which drums should be sampled.
- Select the appropriate sampling device(s) and container(s).
- Develop a sampling plan that includes the number, volume, and locations of the samples to be taken.
- Develop standard operating procedures for opening drums, sampling, sample packaging, and transportation.
- Have a trained health and safety professional determine, based on available information on the waste and site conditions, the appropriate protective equipment and clothing to be used during sampling, decontamination, and packaging of the samples.

When manually sampling from a drum, use the following techniques:

- Keep sampling personnel at a safe distance while drums are being opened. Sample only after completion of opening operations.
- Do not lean over drums to reach drum being sampled, unless absolutely necessary.
- Cover tops with plastic sheeting or other suitable noncontaminated materials to avoid excessive contact with drum tops.
- Never stand on drums. This is extremely dangerous. Use mobile steps or another platform to achieve the height necessary to safely sample from the drum.

- Obtain samples with either glass rods (such as coliwasa or drum thief—for example as shown in Figures 5.15 and 5.16) or vacuum pumps. Do not use contaminated items to sample. The contaminants may contaminate the sample and may not be compatible with the wastes in the drum. Glass rods should be removed prior to pumping to minimize damage to pumps.

Sampling Ponds and Lagoons Ponds and lagoons present both chemical and physical hazards. The potential physical hazards of sampling ponds and lagoons include splash, immersion, and drowning.

Representative sampling procedures usually require five or six samples that include both the aqueous phase and the bottom sludge material. For

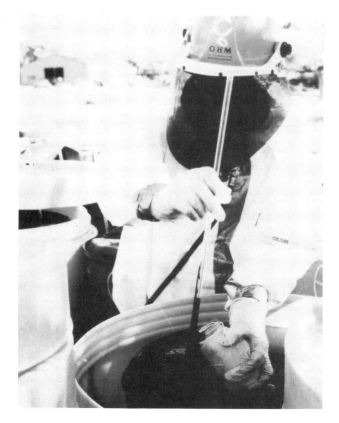

FIGURE 5.15 ■ **A worker uses a glass drum thief to sample drum contents.** *Source:* Courtesy of OHM Remediation Services Corp.

FIGURE 5.16 ■ **Used drum thiefs.** *Source:* Courtesy of OHM Remediation Services Corp.

such sampling, a common but unsafe practice is to lean or reach as far away from the bank as possible, increasing the danger of slipping or falling into the water. Sampling from a boat also presents a hazard. The following safety precautions should thus be observed:

1. Wear protective gear including boots, gloves, and splash goggles.
2. Always sample with two people present.
3. Wear a life preserver or a safety line held by an assistant.
4. A full sample container, particularly at the end of a long pole, will weigh considerably more than when empty. Such an unanticipated strain may cause the person taking samples to overbalance and fall into the pond. If possible, lift samples straight up, using the power of the legs rather than that of the back or arms. If samples are being taken over the side of a boat, the added weight of a full container may be sufficient to cause the boat to tip or rock dangerously. Notify other passengers in the boat when you are about to lift the full sampler. Such warnings will allow passengers to prepare to counterbalance the effects of the weighted sample being lifted. Empty samplers into sample containers in a spill pan, which will catch spillage, drippings, or overflow. Such pans facilitate

cleanup and decontamination. Samplers, sample lines, and related gear should be stored in the pans, rather than in the bottom of the boat where decontamination and cleanup are difficult.

When sampling chemical storage ponds or lagoons, extreme care should be taken to avoid skin contact and immersion. Examples of chemical storage pond or lagoons are shown in Figures 5.17 to 5.19.

5.3 CHEMICAL HAZARDS—OTHER RESOURCE MATERIALS

5.3.1 Material Safety Data Sheets (MSDSs)

Two important sources of information describing the potential hazards of jobsite chemicals are material safety data sheets (MSDSs) and the site safety plan. Every one at the jobsite should know in advance where these references are and the types of information each contains. The usefulness of these will depend on how complete the information is and whether the information is available at the work site.

FIGURE 5.17 ■ Chemical storage pond with nondescript materials floating on top.

152 ■ HAZARD RECOGNITION

FIGURE 5.18 ■ Chemical storage pond with hand-labeled drums in the foreground.

MSDSs are required by the OSHA Hazard Communication Standard (29 CFR 1910.1200). The supplier of the material provides this information to the purchaser. No standard format is required; however, the following points must be covered:

- Contact person at manufacturer/supplier
- Hazardous ingredients/identity information
- Physical data/chemical characteristics
- Fire and explosion hazard
- Reactivity data
- Health hazard data
- Emergency and first aid procedures
- Special protection information
- Special precautions
- Spill, leak, and disposal procedures

The MSDSs along with other resources are important sources of information during preplanning activities. Preplanning must include a review of MSDSs. This will provide information about the properties of each material used on site, how workers should handle the materials, and what personal protective equipment is needed.

5.3 CHEMICAL HAZARDS—OTHER RESOURCE MATERIALS ■ 153

FIGURE 5.19 ■ Two workers collecting samples from a chemical storage lagoon. *Source:* Courtesy of OHM Remediation Services Corp.

Workers must be trained to read MSDSs and to handle the materials that are routinely used. At a manufacturing plant the MSDSs must be available to employees during all shifts. At a remediation site the MSDSs should be included in, or as an appendix to, the site safety plan.

What information does the MSDS provide?
- Preventive measures

   Precautions for safe handling and use

   Storage instructions

   Protective clothing and equipment needed
- Emergency preplanning

   Fire and explosion hazards

   Spill or leak procedures

   Special protection information

154 ■ HAZARD RECOGNITION

Health hazards

First aid information

The supplier/manufacturer should be contacted for additional information as soon as a substance arrives at a work site. The MSDS will include information about how to contact the supplier for information.

Limitations of MSDSs While the MSDS contain important information, there are a number of limitations to their use during an emergency. Some of these limitations are listed below.

1. Limited information

    Information may be incomplete or inaccurate.

    Space on the form may be inadequate.

    Information may not be relevant for the facility.
2. Insufficient time to read the information
3. Insufficient time to call manufacturer/supplier contact
4. Not immediately available

    MSDSs may not be located at the spill or release site.

MATERIAL SAFETY DATA SHEET

Manufacturer:	Address:
Busch Brothers	Busch Lane, Kansas City, MO 63120

Telephone Number for Information:	Emergency Telephone Number:	Date Issued:
316/576-4444	316/576-4441	March 1991

I. Product Identification

Product Name:	Chemical Formula:	C.A.S. Number:

5.3 CHEMICAL HAZARDS—OTHER RESOURCE MATERIALS ■ 155

Ethyl Alcohol 95%	CH_3CH_2OH	64175

Synonyms:	DOT Shipping Description:
Ethanol, 190 Proof Alcohol	Ethyl Alcohol, Flammable Liquid, UN1170

II. Hazardous Ingredients/Identity Information

Component:	Percent:	OSHA PEL:	ACGIH TLV:
Ethyl Alcohol	95.0	1000 ppm (1900 mg/m^3)	1000 ppm (1900 mg/m^3)
Water	5.0	Not Applicable	Not Applicable

III. Physical Data/Chemical Characteristics

Boiling Point:	Specific Gravity (H_2O = 1):	Vapor Pressure:
173°F (78.3°C)	0.816 @ 60°F (15.6°C)	43 mm Hg @ 68°F (20°C)

Percent Volatile by Volume:	Vapor Density (air = 1):	Evaporation Rate(CCl_4 = 1):
99.9	1.6	1.4

Solubility in Water:	Appearance and Odor:
Complete	Clear, colorless, volatile liquid with alcoholic odor

IV. Fire and Explosion Hazard Data

Flash Point:	Flammable/Explosive Limits in Air (% by Volume):	
55°F (12.8°C)	Lower 3.3	Upper 19.0

Autoignition Temperature:	OSHA Class:	Extinguishing Media:
793°F (422°C)	1B	Flammable Liquid Dry Chemical, Alcohol Foam, Water Spray

156 ■ HAZARD RECOGNITION

Special Fire Fighting Procedures:

Do not use ordinary foam.
Do not breath fumes.
Avoid eye and skin contact.
Vapor may travel some distance to a source of ignition and flash back.
Container may rupture violently in fire.

Unusual Fire and Explosion Hazards:

Alcohols can react vigorously with oxidizers.

V. Reactivity Data

Stability: Conditions to Avoid:

X   Stable   Unstable   Keep away from heat, sparks, and flames

Incompatibility (Materials to Avoid):

Capable of reacting vigorously with oxidizing agents, such as nitrates, perchlorates, sulfuric acid, nitric acid, etc.

Hazardous Decomposition Products: Hazardous Polymerization:

Carbon Dioxide   Will not occur

VI. Health Hazard Data

ACGIH Threshold Limit Value: Primary Route(s) of Exposure:

1000 ppm (1900 mg/m^3)   Skin Contact
                                   Skin Absorption
                                   Inhalation

5.3 CHEMICAL HAZARDS—OTHER RESOURCE MATERIALS ■ 157

Signs and Symptoms of Exposure:

Acute: Exposure to ethyl alcohol vapors in excess of 1000 ppm in air may cause headache and irritation of the eyes, nose, and throat. Prolonged exposure may cause symptoms of alcohol intoxication, drowsiness, weakness, loss of appetite, and an inability to concentrate. Exposure to very high concentrations may cause symptoms of alcohol intoxication, headache, drowsiness, tremors, fatigue, dizziness, and unconsciousness. Ingestion of 190 proof alcohol produces the typical effects of alcohol intoxication. Ingestion of very large doses can cause alcohol poisoning and death. Skin contact with liquid ethyl alcohol may cause drying and cracking due to defatting of the tissue.

Chronic: Repeated, prolonged skin contact can cause drying and cracking of the skin and possible dermatitis.

Medical Conditions Generally Aggravated by Exposure:
    X Skin   X Eyes   X Liver   X Respiratory System   Kidneys
    X Central Nervous System

VII. Emergency and First Aid Procedures

Eye Contact:

Flush thoroughly with running water for at least 15 minutes, including under the eyelids.
Get medical attention.

Skin Contact:

Flush area with water.
Remove contaminated clothing.
Get medical attention if irritation persists after flushing.

Inhalation:

Remove to fresh air.
Restore and/or support breathing as required.
Get medical attention.

Ingestion:

If victim is conscious and less than 2 hours have elapsed since ingestion, administer large quantities of water and induce vomiting or gastric lavage. Do not make an unconscious person vomit.
Get medical attention immediately.

VIII. Special Protection Information

Ventilation Requirements:

Local exhaust to maintain concentration of alcohol below 1000 ppm.

Specific Personal Protective Equipment:

Respiratory: NIOSH-approved respirator for organic vapors if ventilation is not adequate. For unknown concentrations, fire fighting, or high concentrations, use a SCBA with full facepiece.
Eye: Face shield, chemical safety goggles, or safety glasses with side shields.
Gloves: Butyl rubber, natural rubber, neoprene, nitrile rubber, polyethylene, Viton, or other compatible materials.
Other: Appropriate protective clothing for the work situation to minimize skin contact.
Eyewash stations and safety showers should be available in areas of handling and use.

IX. Special Precautions

Store in tightly closed containers. Store in cool, dry, well-ventilated, fire-resistant areas.

Store out of direct sunlight, away from oxidizing agents and sources of heat or ignition.

Electrically bond and ground metal containers during liquid transfer.

REVIEW QUESTIONS

To answer the questions in this exercise you may need the following references:

> *DOT Emergency Response Guidebook*
> *Threshold Limit Values*—ACGIH
> *NIOSH Pocket Guide to Chemical Hazards*
> MSDS on ethyl alcohol (see above)

From the information provided in the MSDS for ethyl alcohol and from the above references, complete the following worksheet.

MSDS WORKSHEET

Product Identification

1. Name _____
2. Company or Common Name _____
3. C.A.S. # _____
4. Manufacturer's Name _____
5. DOT Hazard Classification _____
6. DOT Label _____

Components

7. Ingredients _____
8. OSHA PELs _____
9. ACGIH TLVs _____
10. Other Standards _____
11. Is this a toxic or highly toxic substance? _____
12. Is this material considered a carcinogen? _____

Physical Data

13. Boiling Point _____
14. Specific Gravity _____
15. Vapor Pressure _____
16. Vapor Density _____
17. Evaporation Rate _____

18. pH _____
19. Water Solubility _____
20. Explain the importance of items 5 through 19 in relation to spilll response.

Fire and Explosion Data

21. Flash Point _____
22. Explosive/Flammable Limits (LEL/UEL) _____
23. Fire Fighting Media _____
24. Acute Effects of Product Exposure _____
    _____
    _____
25. Chronic Effects _____
    _____
    _____
26. First Aid Recommendations _____
    _____
    _____
27. Primary Routes of Entry _____
28. Other Relevant Information _____

Reactivity Data

29. Stability _____
30. Incompatibilities _____
31. Hazardous Polymerization _____
32. Explain the importance of items 21–31 in relation to spill response.
    _____
    _____
    _____
    _____

Spills and Leaks

33. What absorbent media can be used? _____
    _____
    _____

Personal Protective Equipment

34. What personal protective equipment should be used for spill response? _____

_____

_____

35. What is the reportable quantity? _____

REFERENCES

American Conference of Governmental Industrial Hygienists, *Threshold Limit Values for Chemical Substances and Physical Agents and Biological Exposure Indices with Intended Changes for 1994–1995*, ACGIH, Cincinnati, OH, 1995.

Cote, A. E., ed., *Fire Protection Handbook*, 16th ed., National Fire Protection Asssociation, Quincy, MA, 1986.

DOT 49 CFR171-173. *Hazardous Materials Transportation Act*, U.S. Government Printing Office, Washington, D.C.; 1991.

EPA, *Compendium of ERT Waste Sampling Procedures*, U.S. Government Printing Office, Washington, D.C., 1991.

Levine, S., and W. Martin, eds., *Protecting Personnel at Hazardous Waste Sites*, Butterworth, Stoneham, MA, 1985.

NIOSH/OSHA/USCG/EPA, *Safety and Health Guidance Manual for Hazardous Waste Site Activities*, U.S. Government Printing Office, Washington, D.C., 1985.

O'Connor, C. J., and S. I. Litzman, *Handbook of Chemical Industry Labeling*, Noyes, Park Ridge, NJ, 1984.

OSHA 29 CFR 1910.1200: *Hazard Communication*, U.S. Government Printing Office, Washington, D.C., 1983.

U.S. Department of Health and Human Services, *NIOSH Pocket Guide to Chemical Hazards*, U.S. Government Printing Office, Washington, D.C., 1994.

U.S. Department of Transportation, *1990 Emergency Response Guidebook*, DOT P 5800.5, Washington, D.C., 1990.

Wagner, K., R. Wetzel, C. Bryson, A. Wickline, and V. Hodge, *Drum Handling Manual for Hazardous Waste Sites*, Noyes, Park Ridge, NJ, 1987.

Waxman, Michael, F., ed., *Emergency Response Team Training for Technicians—24-Hour OSHA Training Workshop*, University of Wisconsin, Madison, WI, 1993.

Waxman, M. F., and D. W. Kammel, *A Guidebook for the Safe Use of Hazardous Agricultural Farm Chemicals and Pesticides*, U.S. Department of Agriculture, North Central Regional Cooperative Extension Publications 402, Madison, WI, 1991.

CHAPTER 6

PERSONAL PROTECTIVE EQUIPMENT

CHAPTER OBJECTIVES

When you have completed this chapter, you will be better able to

- Describe the general types, uses, and limitations of chemical protective clothing (CPC)
- Identify the CPC required for the various EPA levels of protection
- Select the most appropriate CPC to mitigate the effects of potential exposure to chemicals on-site
- Don and doff CPC in the appropriate manner and describe the precautions to be taken while suited up
- Properly inspect, maintain, and store CPC

6.1 INTRODUCTION

Anyone entering an uncontrolled hazardous waste site or responding to a hazardous chemical emergency must be protected against any potential hazards they may encounter. The purpose of personal protective clothing and equipment is to shield or isolate individuals from the chemical, physical, and biological hazards presented by the site or incident. Careful selection of adequate personal protective equipment (PPE) should protect

- Respiratory system
- Skin and body
- Face and eyes
- Feet and hands
- Head
- Hearing

This chapter describes the various types of PPE that are appropriate for use on uncontrolled hazardous waste sites and during emergency response to chemical incidents. It also provides guidance in their selection and use. Chapter 7 will cover respiratory protection.

More detailed guidance on chemical protective clothing (CPC) selection and use can be found in *Guidelines for the Selection of Chemical Protective Clothing*, published by the ACGIH (Schwope et al., 1987), and *Quick Selection Guide to Chemical Protective Clothing*, published by Van Nostrand Reinhold (Forsberg and Mansdorf, 1993).

6.2 REQUIREMENTS FOR THE USE OF PERSONAL PROTECTIVE EQUIPMENT

Use of personal protective equipment (which includes CPC) is required by OSHA regulations in 29 CFR 1910 (see Table 6.1) and reinforced by EPA regulation 40 CFR 300.71, which requires all private contractors working on Superfund sites to conform to applicable OSHA provisions and any other federal or state requirements deemed necessary by the lead agency overseeing the activities. Additional respiratory regulations, passed pursuant to the Mine Safety Act, are found in 30 CFR Part 11.

No one piece of protective equipment, nor any single combination of equipment and clothing, is capable of protecting against all potential hazards. In fact, no protective equipment is capable of providing protection for a prolonged period of time against even one chemical. Thus PPE should be considered the last option for protection, used only after other safety procedures, alternate remedial actions, and/or engineering controls have been considered.

The use of PPE itself creates significant worker hazards, such as heat stress, physiological stress, impaired visability, mobility, and communication.

TABLE 6.1 ■ OSHA Regulations and Sources for Use of Personal Protective Clothing and Equipment

Type of Protection	Regulation	Source
General requirements	29 CFR 1910.132	41 CFR 50-204.7: *General Requirements for PPE*
Eye and face	29 CFR 1910.133(a)	ANSI Z87.1-1968: *Eye and Face Protection*
Respiratory protection	29 CFR 1910.134	ANSI Z88.2-1969: *Standard Practice for Respiratory Protection*
Head protection	29 CFR 1910.135	ANSI Z89.1-1969: *Safety Requirements for Industrial Head Protection*
Foot protection	29 CFR 1910.136	ANSI Z41.1-1967: *Men's Safety Toe Footwear*
Hearing protection	29 CFR 1910.95	

Source: NIOSH/OSHA/USCG/EPA, *Occupational Safety and Health Guidance Manual for Hazardous Waste Site Activities*, 1985.

The greater the level of PPE protection, the greater are the associated risks. For any given situation, equipment and clothing should be selected that provide an adequate level of protection. Overprotection may be hazardous and should be avoided.

6.3 SELECTION OF PPE FOR SITE ENTRY

When workers use any article or ensemble of CPC on a hazardous waste site, the advantages and limitations of the equipment should be carefully considered in regard to the potential exposures to chemical and physical hazards. Consequently, selection of the equipment should be performed by individuals who are familiar with both the equipment and the likely conditions under which the PPE will be used.

Frequently at hazardous waste sites, the chemicals and mixtures of chemicals are unknown; often, the visible, physical characteristic of the chemicals (solid, gas, liquid) or the odor are the only available information. Thus it is difficult to assess the degree of hazard to which workers may be exposed. The safest, most conservative and recommended approach for selecting PPE in such cases is to assume initially the worst exposure condition and use the highest level of PPE. Then, as the hazardous agents on site become characterized, the PPE can be adjusted to match the specific hazards. A good example of this approach is described by the EPA advisory protocol for hazardous waste site entry (see Table 6.2).

The left side of Table 6.2 lists the selection criteria under the level of protection. If any of the selection criteria listed under Level A were present on site, then Level A PPE must be used. As the concentration of hazardous substances, potential for splash, and organic vapor levels are reduced, the level of PPE can be lowered to Level C. Level D PPE is primarily a work uniform and should not be worn where there is a potential for exposure through the skin or foot or by inhalation.

The selection of clothing requires an evaluation of the hazards at the site, the chemicals present, the potential routes of entry into the body by the hazardous chemicals, and a knowledge of, or sources to obtain information regarding, the most effective chemically resistant materials of construction. The common, recommended practice is to use the minimum necessary amount of PPE to provide protection to the worker. PPE can be burdensome and restrictive; minimizing PPE increases the likelihood that it will be worn. This also minimizes the loss in worker efficiency that typically accompanies PPE use. In addition, PPE can be expensive and it must be either decontaminated and properly maintained or disposed of after use.

The determination of the type of clothing that is appropriate for a given job is the responsibility of the industrial hygienist or occupational safety officer. These professionals must consider in their decision-making process all aspects of the job activities to be performed, the conditions under which they will be done, and the capabilities of the workers.

6.4 DEFINITIONS

Chemical Resistance: The ability of a material to stop a chemical from passing through it. A chemical can get through a material and into a garment in three ways. Knowing what these ways are and how they affect the CPC are keys to CPC selection and use. Chemicals can pass through a material by

1. *Penetration:* The movement of material through a hole in the CPC. Holes such as rips or tears, can be accidental, or they can be part of the con-

TABLE 6.2 ■ Environmental Protection Agency Site Entry Protocol

Level of Protection/Selection Criteria	Full Body	Respiratory Protection	Hands	Feet	Eyes/Face	Head
Level A						
Chemicals and concentrations unknown/above safe level Extremely hazardous substances/skin destructive	Fully encapsulated suit	SCBA (pressure demand)/Airline	Two pairs of gloves	Chemical-resistant steel toe	Full facepiece	Hard hat
Level B						
IDLH or concentrations above PF of APR/less than 19.5% O_2. Unidentified gases or vapors	One- or two-piece chemical-resistant suit with hood	SCBA (pressure demand)/Airline	Two pairs of gloves	Chemical-resistant steel toe	Full facepiece	Hard hat
Level C						
Contaminants and chemical concentration known/APR OK. No IDLH/O_2 above 19.5%	One- or two-piece suit/disposable suit	APR half or full facepiece	Two pairs of gloves	Steel-toe safety boots	Goggles or full facepiece APR	Hard hat
Level D						
No measurable chemical concentrations. No inhalation or splash exposure potential	Coveralls	None	One pair of gloves	Steel-toe safety boots	Safety glasses	Hard hat

Source: EPA, *Standard Operating Safety Guides*, 1984.

struction, such as needle holes from sewing the suit together or from the use of non-chemical-resistant zippers.
2. *Degradation:* The breakdown of one or more of the properties of a material. The material may harden, weaken, and/or stretch; it may dissolve and disappear; or, if the liquid is hot, it may melt.
3. *Permeation:* The movement of a chemical through the material of the chemical-protective suit. CPC will not keep all substances away from the wearer's skin for an indefinite period of time. Many chemicals slowly permeate (work their way through) the clothing, even the fully encapsulating suits. Chemicals that have permeated a material are difficult or impossible to detect and remove. If the contaminants are not removed during the decontamination process, they may cause an unexpected exposure to the next wearer. Suits that cannot be decontaminated must be disposed of properly.

Manufacturers test or have their CPC tested for chemical resistance; upon request, they normally supply these data with the clothing. Table 6.3 provides an example of how these data are helpful. Which suit would you use?

6.5 SELECTION CRITERIA FOR DETERMINING THE PROPER CPC

Temperature Resistance: The ability of a material to resist changes in its chemical composition and/or physical properties when exposed to extremes of temperature. The effect on the material can range from becoming brittle and breaking in cold, to stretching and melting in high temperatures and upon exposure to hot chemicals. Workers must take

TABLE 6.3 ■ Breakthrough Time (Hours)

Chemical	Butyl Rubber	Neoprene	Nitrile	Polyvinyl Alcohol	Polyvinyl Chloride	Viton
Acetone	20.33	0.17	0.33	>4.00	0.30	0.01
Sulfuric Acid	>8.00	2.50	>6.00	NR	1.75	>8.00
Tetrachloroethylene	0.17	0.12	4.00	5.00	0.01	>17.00
Toluene	0.17	0.03	1.00	>25.00	<0.01	>16.00

Source: Taken, in part, from *Guidelines for the Selection of Chemical Protective Clothing*, 3rd ed., Environmental Protection Agency, 1987.

care when handling (a) extremely cold liquids such as liquid chlorine at temperatures as low as $-212°F$ and (b) chemicals that upon mixing result in an exothermic reaction; for example, mixing sulfuric acid and water will produce such a reaction.

Physical Integrity: Construction of the suit is important for proper function. Suits that rip or have seams that pull apart, or suits improperly sewn, offer little protection. Sealed seams and protected zippers minimize penetration through them.

Cost: Initial and ongoing cost of purchasing CPC can be an important consideration for management. When shopping prices, be sure the lower price is for garments of equal quality. Such quality considerations include the materials of construction, the type of seams, faceshield material, and the type of zipper and boots.

Shelf Life: Most CPC now have a specified shelf life. If they are not used by this date, they must be removed from service. This also will require routine replacement of equipment.

Size: CPC should be available in a variety of sizes to accommodate the height and weight of site workers. Suits that are too small will tear easily and provide no protection. Suits that are too large will make working difficult.

Glove Selection Factors that determine the type of gloves used are the materials of construction, type of construction, and length.

1. **Materials of Construction.** Many types of plastic and elastomeric compounds and formulations are used by manufacturers. Request chemical resistance test results for gloves under consideration for purchase from the manufacturer or vendor.

2. **Type of Construction.** Two common types of construction are used:

   Supported. These have a cloth lining that has been partly or fully coated with chemical resistant material.

   Unsupported. These are formed on a hand shaped mold by dipping the mold into the chemical resistant material until the it builds to a specified thickness then removed form the mold. These gloves have no cloth lining inside.

3. **Length.** Gloves are available in several lengths that vary from wrist length to over the elbow.

6.5.1 Summary: Selection Criteria for CPC
Considerations for Selection of CPC

Chemical Resistance. Different materials are resistant to different chemicals. Management should provide CPC which will provide protection against the chemicals likely to be encountered during emergency response at the plant.

Physical Integrity. Construction of the suit is important for the proper functioning of the CPC. Seams and zippers should provide solid barriers to chemicals as well as be constructed in a manner which provides some flexibility to allow the wearer free movement.

Resistance to Temperature Extremes. Heat and cold can adversely affect CPC. Clothing which will be worn in cold temperatures could crack or become ineffective against chemicals. Likewise, heat may destroy the chemical resistance of clothing or even melt the clothing.

Ability to Clean. Clothing must be able to be cleaned and decontaminated after each use. If this is not possible, the clothing must be disposed of after use.

Cost. Initial and ongoing costs of purchasing CPC can be an important consideration for plant management. However, buying less expensive, inferior products which do not adequately protect employees can be more expensive, in the long run, due to loss of human life or lost work time and medical costs.

Flexibility. Materials need to be flexible for the wearer to move and work safely. Overly rigid suits can result in unnecessary accidents from slips, trips, and falls.

Size. CPC should be available in a variety of sizes to accommodate the height and weight of the responders. Suits that are too small will tear easily and provide no protection. Suits that are too large will make walking and/or working difficult.

Breakthrough time is a measure of the time in hours it takes for a chemical from the onset of direct contact with the resistant material until the chemical permeates the material. Materials that provide no resistance are indicated by NR, not recommended (see sulfuric acid: polyvinyl alcohol, PVA, is water-soluble; therefore, aqueous solutions such as acids should not be used with PVA). Manufacturers and vendors of CPC will supply such data upon request. Use only the data supplied by the manufacturer of the clothing you plan to use. Resistant materials vary from manufacturer to manufacturer; because material thickness varies, make sure you are using the appropriate permeation data.

6.6 LEVELS OF PROTECTION—EPA DEFINITIONS

The EPA defines four levels of personal protective equipment. OSHA has adopted the same system. The level selected will depend on the degree of protection needed for the emergency.

Level A provides the highest level of skin, respiratory and eye protection.

Level B provides the highest level of respiratory protection but a lower level of skin protection.

Level C provides a lower level of skin, respiratory and eye protection and should only be used when working with specific known substances at appropriate concentrations.

Level D provides minimal protection from chemical hazards.

The employer should describe in the emergency response plan the level of PPE that will be used for specific types of emergency situations.

When Is Level A Protection Needed? Level A protection is required when the highest level of protection for skin, eyes, and the respiratory system is needed.

- When a SCBA is required
- When hazardous substances may be harmful to skin
- When hazardous substances may be absorbed through skin
- When there is potential for splash or immersion in a liquid
- When confined space entry may be involved and the need for Level A cannot be ruled out (but explosion hazard has been ruled out)
- When the oxygen level or toxic concentrations are not known or near IDLH

What Is Level A Protection? Level A protection *must* include the following equipment (see Figure 6.1):

- Positive-pressure, full-facepiece SCBA or positive-pressure supplied-air respirator with escape SCBA

172 ■ PERSONAL PROTECTIVE EQUIPMENT

FIGURE 6.1 ■ **Worker dressed in Level A protection.** *Source:* Courtesy of MSA (Mine Safety Appliances Co.).

- Chemical-resistant, inner and outer gloves
- Chemical-resistant, encapsulating, protective suit
- Chemical-resistant boots with steel toe and shank

The following equipment is *optional* Level A equipment and is to be used as applicable and/or appropriate.

- Coveralls
- Long underwear
- Hard hat under suit
- Disposable protective suit, gloves and boots (may be worn over suit for added protection, suit protection, and/or ease of decontamination)

When Is Level B Protection Needed? Level B protection is required when the highest level of respiratory protection is needed but a lower level of skin protection is acceptable.

- When the type of substances have been identified
- When supplied air is required
- Less skin protection is needed (vapor and gases are not believed to contain high levels of chemicals harmful to skin or capable of being absorbed through intact skin)

What Is Level B Protection? Level B protection must include the following (see Figures 6.2 and 6.3):

- Positive pressure, full-facepiece SCBA or positive-pressure supplied-air respirator with escape SCBA (NIOSH approved)
- Hooded chemical-resistant clothing
- Inner and outer chemical-resistant gloves
- Outer chemical-resistant boots with steel toe and shank

The following are optional Level B equipment and are to be used as applicable and/or appropriate.

FIGURE 6.2 ■ **Two workers attired in Level B using SCBA for respiratory protection.** *Source:* Courtesy of MSA (Mine Safety Appliances Co.).

174 ■ PERSONAL PROTECTIVE EQUIPMENT

FIGURE 6.3 ■ **Worker dressed in Level B protection using an air-line respirator.** *Source:* Courtesy of MSA (Mine Safety Appliances Co.).

- Coveralls under CPC
- Outer, chemical-resistant boot covers (disposables may be used to aid decon)
- Hard hat
- Face shield

When Is Level C Protection Needed? Level C provides protection when the concentrations and types of airborne substances are known and the criteria for using an air-purifying respirator are met.

- When direct contact with the hazardous substance will not harm the skin or be absorbed through any exposed skin
- When air contaminants have been identified, concentrations measured, and an air-purifying respirator is available that can remove the contaminants
- When an adequate level of oxygen (>19.5%) is available, and all other criteria for the use of air-purifying respirators are met

6.6 LEVELS OF PROTECTION—EPA DEFINITIONS

What is Level C Protection? Level C protection must include (see Figure 6.4):

- Full-face or half-mask air-purifying respirator (NIOSH approved)
- Hooded chemical-resistant clothing
- Inner and out chemical-resistant gloves

The following are optional Level C equipment and are to be used as applicable and/or appropriate:

- Coveralls
- Outer chemical-resistant boots with steel toes and shank
- Chemical-resistant boot covers
- Hard hat
- Escape mask
- Face shield

When Is Level D Protection Needed? Level D is required when minimal protection from chemical exposure is needed. It is worn to prevent nuisance contamination only when:

FIGURE 6.4 ■ Workers dressed in Level C protection with APR.

- The atmosphere contains no known hazards
- Work functions preclude splashes, immersion, or the potential for unexpected inhalation of, or contact, with hazardous levels of any chemicals.

What Is Level D Protection?

- Level D protection must include the following equipment (see Figure 6.5).
- Coveralls (work uniform)
- Chemical-resistant boots or shoes with steel toe and shank
- Hard hat

The following are optional Level D equipment and are used as applicable and/or appropriate:

- Gloves
- Outer, chemical-resistant boots (disposable)
- Safety glasses or chemical splash goggles

FIGURE 6.5 ■ **Worker monitoring soil contamination dressed in Level D.** *Source:* Courtesy of MSA (Mine Safety Appliances Co.).

- Escape mask
- Face shield

A general rule for which level of protection to use is: *"The less you know, the higher you go."*

A helpful way to remember the levels of protection is:

Level A—"A"ll covered

Level B—"B"reathing air

Level C—"C"artridge respirator

6.7 PRECAUTIONS WHEN WEARING CPC

Always use the *buddy system* when donning/doffing and wearing CPC. Buddies should watch for signs of equipment failure and fatigue and assist in donning/doffing.

Materials used to make most suits do not "breathe." Rapid heat and moisture buildup will occur in the suit during use. Look for signs of *heat stress* (dizziness, headache, nausea, sweating stops) especially at temperatures over 70°F.

Due to size and weight of suits, *motion is restricted* especially when climbing, working in tight areas, or using hand tools.

Encapsulating suits restrict *vision*. Use caution when working in them.

Seams are the weak point of suits, especially disposable ones. Use caution not to strain and split them, If this occurs, report the incident and follow the appropriate SOP (standard operating procedure).

Whenever possible, a *variety of suit sizes* should be on hand to fit the various sizes of team members. If a suit being worn is too large, it can be adjusted by pulling up excess and taping or using an adjustable harness inside the suit. *Caution:* Some suits should not be taped because the adhesive can degrade the suit.

Suits offer *no fire protection* and, in some cases, increase the possibility of injury because they will melt. Use caution when suits are used in potential fire areas.

All suits have limits as to the *temperature* in which they can be worn without damage. Check the manufacturer's data.

Some people are *claustrophobic* and are unable to wear encapsulated suits. It is important to identify people who are claustrophobic and determine before an emergency occurs if they will be able to wear the encapsulated suit. It they cannot tolerate the suit, they should not be expected to respond to an emergency where an encapsulated suit is required.

When wearing fully encapsulated suits, the wearer must be able to access the emergency controls of the SCBA. Preferably, the individual should be able to pull his/her hand into the main part of the suit to access the controls of the SCBA.

Disposable booties may be slippery. Use caution when walking to prevent slips and falls.

Hearing and speaking is difficult in PPE. It is important to establish other ways to communicate with each other. Hand signals or audio signals such as horns, sirens, and whistles can be used to communicate (see Figure 6.6). Communication can also be improved by using two-way radios, such as (a) a portable radio with a microphone or (b) a radio with a microphone and speaker combination, attached to the full-face respirator.

Caution: Always use the buddy system. Know how to communicate.

6.7.1 Permeation

Chemicals can permeate or go through the material of the chemical-protective suit. CPC will not keep all substances away from the wearer's skin for an indefinite period of time.

Many chemicals slowly permeate (work their way through) the clothing, even the fully encapsulation suits. Chemicals that have permeated a material are difficult or impossible to detect and remove. If the contaminants

Everything is Fine   Something is Wrong

FIGURE 6.6 ■ Hand signals used to communicate when ambient noise levels are "too high" to communicate verbally.

are not removed during the decontamination process, they may cause an unexpected exposure to the next wearer. Suits which cannot be decontaminated must be disposed of properly.

The rate of permeation is dependent on five major factors:

Contact Time. The longer a contaminant is in contact with an object, the greater the probability and extent of permeation. (Therfore, minimizing contact time is one of the most important objectives of a response activity and decontamination program.)

Concentration. As the concentration of chemicals increases on the outside of the CPC, the greater the potential for permeation.

Temperature. As the temperature increases, the permeation rate of contaminants generally increases.

Type of Chemical. Some chemicals permeate CPC more readily than others.

Physical State. As a rule, gases, vapors, and "runny" liquids tend to permeate more readily than "thick" liquids or solids.

Note: Data are not available for all chemicals because manufacturers tend to test against what they know the suits to be good against. If data are not available for your type of chemicals, consult the manufacturer or an independent test results database.

NFPA Standard 1992 for CPC requires that all suits be tested against 17 standard chemicals, and results are independently certified by the Safety Equipment Institute (SEI). Look for these results to compare the resistance of the suits.

6.8 INSPECTION, MAINTENANCE, AND STORAGE OF CPC

It is important to inspect CPC to detect any evidence of chemical breakthrough. CPC which is torn, degraded, or otherwise malfunctional will not offer adequate protection to the wearer. The ERP should describe or reference SOPs for inspection, maintenance, and storage of CPC.

CPC should always be inspected when it is

- Received from the distributor
- Issued to workers

- Stored
- Taken out of storage
- Used for training
- Used for work or an emergency response
- Returned from maintenance

Note: Some CPC has a "shelf-life" and should be discarded when the expiration date passes.

An inspection checklist should be developed for each item. Factors to consider are

- Cuts, holes, tears, and abrasions in seams of fabric
- Weakness in zipper or valve seals
- Signs of incomplete decontamination (discolorations)
- Signs of malfunctioning exhaust valves

Note: Do not assume that if the CPC does not appear discolored, that it is decontaminated.

Proper maintenance can correct CPC deficiencies and prolong the CPC's life. A detailed SOP should be developed and followed rigorously.

Proper storage of CPC is important to prevent suit failures. A written SOP should describe storage before PPE is issued to the wearer (in a warehouse, on-site, etc.) and storage of PPE after usage.

6.9 SUMMARY

PPE includes respirators, chemical-resistant suits, boots, gloves, chemical goggles, and face shields. PPE is required by OSHA regulations and protects emergency responders from

- Chemical contact with skin and eye
- Temperature
- Respiratory hazards

Levels of PPE *Level A* provides the most protection and includes

- A positive-pressure, full-facepiece SCBA or supplied air with escape unit.
- A totally encapsulating chemical-resistant suit
- Inner and outer chemical-resistant gloves
- Chemical-resistant boots with steel toe and shank

Level B includes

- A positive pressure, full-facepiece SCBA or supplied air with escape unit
- Hooded, chemical-resistant clothing
- Inner and outer chemical-resistant gloves
- Chemical-resistant boots with steel toe and shank

Level C includes

- Full- or half-face air-purifying respirator (APR)
- Hooded, chemical-resistant clothing
- Inner and outer chemical-resistant gloves
- Chemical-resistant boots with steel toe and shank

Level D includes

- Hard hat
- Coveralls
- Chemical-resistant boots with steel toe and shank

PPE must be properly cared for and maintained. Wearers should know the requirements of PPE. Written programs about selection, care, and use of PPE should be included in or referenced in the Emergency Response Plan.

Things to Remember When Using Chemical-Protective Clothing

- No outfit protects against all hazards
- The clothing will make it difficult to move.
- Practicing work tasks while wearing PPE will help prepare for an emergency.
- Watch out for the signs of heart stress (dizziness, headache, nausea, sweating stops) while working in protective clothing.
- Always work with a buddy; know how to communicate.
- Proper inspection, cleaning, storage, and maintenance are necessary to ensure that PPE will protect the responder.
- Tape may degrade the clothing.
- If clothing is torn or damaged, report it immediately to the appropriate person.
- Suits must be selected appropriately for the hazard.
- PPE must always be properly decontaminated.

REVIEW QUESTIONS

1. CPC is selected based on what considerations?
2. List three situations at a hazardous waste site that could require CPC to prevent potential exposures? What level of protection (A,B,C) is required for each?
3. List some precautions to take while wearing CPC.
4. When should PPE be inspected?
5. When should PPE be replaced?
6. The following exercise will allow you to apply knowledge gained from this section to a "real-life" situation. The exercise involves determining what level of PPE would be required for different situations. For each situation, state the appropriate level of PPE and the reason for your decision.
    a. At a loading dock at the site at which you are working, a truck has leaked unknown materials onto the ground. The material is vaporizing and you do not have any monitoring equipment. What level of protection should you wear to size-up the scene near the release?
    b. A drum containing a mixture of trichloroethylene (TCE) and other spent solvents has a minor leak. The TCE level is measured at about 900 ppm (1000 ppm is IDLH for TCE). What level of PPE should you wear to patch the drum and overpack it?
    c. Leaking drums are reported in a test pit. The oxygen concentration is approximately 18%. The combustible gas indicator reads 45% of the LEL. What CPC should you wear to enter the pit and sample the drums?
    d. An underground storage tank that previouly contained leaded gasoline has apparently leaked. The tank is being excavated and the contaminated soil is now visible. What level of protection do you require to sample the contaminated soil?

REFERENCES

EPA, *Standard Operating Safety Guides*, U.S. Government Printing Office, Washington, D.C., 1984.

Forsberg, K., and S. Z. Mansdorf, *Quick Selection Guide to Chemical Protective Clothing*, 2nd ed., Van Nostrand Reinhold, New York, 1993.

NIOSH/OSHA/USCG/EPA, *Safety and Health Guidance Manual for Hazardous Waste Site Activities*, U.S. Government Printing Office, Washington, D.C., 1985.

Schwope, A. D., P. P. Costas, J. O. Jackson, J. O. Stull, and D. J. Weitzman, *Guidelines for the Selection of Chemical Protective Clothing*, ACGIH, Cincinnati, OH, 1987.

CHAPTER 7

RESPIRATORY PROTECTION

Respiratory protection is vital for the site worker and the first responder while they are engaged in activities with hazardous substances. This chapter describes the various types of respiratory protection appropriate for use at uncontrolled hazardous waste sites and during emergency response to chemical incidents. It also provides guidance in their selection and use.

CHAPTER OBJECTIVES

When you have completed this chapter, you will be able to

- Properly test the fit of your respirator facepiece
- Identify situations where respiratory protection is needed
- Select the proper type of respirator to provide the protection required
- Clean and maintain your respiratory equipment
- Describe the key components of a respiratory protection program

7.1 INTRODUCTION

Respiratory protective devices vary in design, equipment specifications, applications, and protective capability. Proper selection depends on the toxic substance involved, conditions of exposure, human capabilities, and equipment fit.

Providing respiratory protection to site workers and responders is by far the most important aspect of protective equipment. Toxic materials have three entry routes into the body:

- Inhalation
- Skin absorption
- Ingestion

Inhalation presents the quickest and most direct route into the body and to the bloodstream. The level of protection which can be provided ranges from a single-use air-purifying respirator to a positive-pressure demand self-contained breathing apparatus (SCBA). Therefore, the proper selection of a respirator for an specific hazard becomes a systematic evaluation.

The two basic respiratory hazards encountered are oxygen deficiency and contaminated atmospheres. The normal oxygen content in the atmosphere is 20.9% by volume. Oxygen content below 10% will not support combustion and is considered unsafe. At low oxygen concentrations an individual can collapse immediately without warning, and death can result in minutes. While 16% (at sea level) is considered the lowest level safe, current legislation and industry standards require that a work area not have less than 19.5% oxygen. When exposure conditions are being assessed, it is important to remember that oxygen deficiency can occur in enclosed or confined spaces where the oxygen is being displaced by other gases or by vapors, or by fire, hot work, and rust where the oxygen is being consumed.

Air contaminants include particulate solids or liquids, gaseous material in the form of a true gas or vapor, or a combination of gas and particulate matter. The type of respiratory hazard posed to the individual dictates the particular respirator to be used.

7.2 RESPIRATORY PROTECTION—SELECTION CRITERIA

7.2.1 Airborne Contaminants

Particulate Hazards Particulate hazards may be classified according to their chemical and physical properties and their effect on the body. Particle diameter in micrometers (μm) is one of the most important properties. Particles below 10 μm in diameter have a better opportunity to enter the respiratory tract, and particles in the range of 1–2 μm in size can reach the deep lung spaces. A healthy lung will generally clean out the particles in 5- to 1-μm range because these particles remain in the upper airways. With increased exposures or diseased systems, the efficiency of the lung is reduced.

The types of particulate hazards are classified as follows:

- *Dust*. Mechanically generated solid particulate (0.5–10 μm)
- *Mist and fog*. Liquid particulate matter (5–100 μm)
- *Fumes*. Solids condensation particles of small diameter (0.1–1.0 μm)
- *Smoke*. Chemically generated particulate (solid and liquid) of organic origins (0.01–0.3 μm)

Gaseous Contaminants Gaseous contaminants can also be classified according to their chemical properties:

- Inert gases (helium, argon, etc.), which do not metabolize in the body but displace air to produce an oxygen deficiency
- Acid gases (SO_2, H_2S, HCl, etc.), which are acids or produce acid reactions with water
- Alkaline gases (NH_3, etc.), which are alkalis or produce alkalis by reaction with water
- Organic gases, which exist as true gases or vapors from organic liquids
- Organometallic gases (metals attached to organic groups such as tetraethyl lead and the organic phosphates)

Exposure Levels The degree of effect of both gaseous and particulate hazards depends mostly on the airborne concentration of contaminants and the length of exposure. Therefore, a proper assessment of the hazards involved with the particular activity, be it a remedial activity or an emergency response, is the first step in protecting you and your coworkers. This assessment involves the identification of the substance or substances involved and determining the concentration of the each. In order to assess the hazards, air monitoring and sampling is performed to determine the identity and concentration of the chemical contaminants. Sampling and monitoring activities should be performed in areas where the highest concentration of the chemical would be expected. The persons assigned the task of the initial monitoring and sampling should be in the highest level of protection available. After the monitoring results and samples have been collected and analyzed, decision on hazard and level of protection can be confidently made. The next step is the selection of respiratory devices.

7.2.2 Respirator Selection

Protection Factors Protection factors are a measure of the overall effectiveness of a respirator. These numbers have been assigned to an entire class of respirators. The protection factors used by NIOSH and ANSI (Z88.2-1980) are based on quantitative fit tests.

Protection factors (PFs) are determined by dividing the ambient airborne concentration by the concentration inside the facepiece. Instead of establishing protection factors based on quantitative fit testing, today's emphasis is upon sampling in the workplace under workplace conditions. As testing data are updated, the protection factors are likely to change.

$$PF = \frac{\text{Particulate aerosol count outside the respirator}}{\text{Particulate aerosol count inside the respirator facepiece}}$$

The *maximum use limitation* of a APR is that concentration above which the respirator will not provide the necessary protection to the user; therefore, another respirator should be selected that will provide the required protection.

$$MUL = PF \times TLV$$

Example: What is the maximum use limitation for a half-mask air-purifying respirator with organic vapor cartridges in an atmosphere containing perchloroethylene?

The TLV for perchloroethylene (tetrachloroethylene) is currently 25 ppm. The PF for a half-mask respirator is 10 (see Table 7.1). Therefore,

$$MUL = TLV \times PF = 25 \times 10 = 250 \text{ ppm}$$

The maximum use concentration (MUC) for all organic vapor cartridges is 1000 ppm, and therefore we can use this half-mask APR up to a concentration of 250 ppm. If we were working under conditions with ambient concentrations above 250 ppm perchlorethylene, we would have to upgrade our respiratory protection a full-facepiece APR or to a supplied-air respirator.

TABLE 7.1 ■ Recommended Protection Factors[a]

Respirator	ANSI Recommended Protection Factor
Half-mask APR	10
Full-facepiece APR	50
Powered APR (PAPR)	50
Airline respirator-pressure demand with egress	10,000+
Self-contained breathing apparatus-pressure demand (SCBA)	10,000+

[a]There are differences between the NIOSH, OSHA, and ANSI assigned protection factors. The recommendations listed in the table are from ANSI, which are the most conservative values of the organizations listed in the text.

When using a supplied-air respirator (PF = 10,000+) recommended for hazardous materials activities, the MUL for this respirator in an environment containing perchloroethylene would be calculated as follows;

$$MUL = TLV \times PF = 25 \times 10,000 = 250,000 \text{ ppm}$$

These results indicate that a supplied-air respirator, either an airline with egress provisions or an SCBA, would more than be adequate for these conditions.

Both the PF and MUL calculations are accurate only when the fit of the respirator is adequate as determined by a fit test (quantitative and/or qualitative).

7.2.3 Respirator Fit-Testing

After consideration of details pertaining to respirator selection, proper protection will not be provided if the respirator facepiece does not fit the wearer properly. Because of the great variety in face sizes and shapes encountered in the male and female workers, most respirator manufacturers make their models of respirators available in more than one size. In addition, the size and shape of the facepiece varies among the different manufacturers. In other words, the medium-size half-mask facepiece of one manufacturer is not the same shape and size as the medium-size half-mask facepiece from another manufacturer. For these reasons, your respirator fit-test program should make allowances for this variability by providing several commercially available respirators in different sizes when employees are fit-tested.

The OSHA Standard, 29 CFR 1910.134, requires that all negative-pressure respirators (APRs) be fit-tested by exposure to a "test atmosphere." This can be achieved by one of two fitting methods, qualitative or quantitative fit-testing.

Routine Tests Two types of testing, positive- and negative-pressure tests, should be done each time a respirator is donned. These tests can be done in the field to check the seal of the respirator. They do not replace yearly fitting, but provide a routine assessment as to whether or not the fit is still adequate.

Positive pressure test

Purpose:	Checks the apparatus for leaks at valves or other points.
Method:	Wearer covers the exhalation valve with hand and gently exhales. If the respirator has been properly donned, a slight positive pressure can be built up inside the facepiece without the detection of any outward leakage of air between the sealing surface of the facepiece and the wearer's face.
Requirements:	Should be performed prior to each use.

Negative-Pressure Test

Purpose:	Checks the facepiece-to-face seal.
Method:	SCBA wearer places hands over the intake valves and inhales. APR wearer places hands over cartridges and gently inhales and holds his or her breath for at least 10 seconds and exhales. No outside air should be felt leaking into the facepiece. Figure 7.1 shows a worker performing a negative pressure test.
Requirements:	Should be done prior to each use.

Positive- and negative-pressure tests can be done quickly and easily in the field. They do, however, have the disadvantage of relying on the wearer's ability to detect the leaks.

7.2 RESPIRATORY PROTECTION—SELECTION CRITERIA ■ 191

FIGURE 7.1 ■ Trainee performing negative pressure test.

Qualitative Fit-Testing A qualitative fit-test relies on the wearer's subjective response to a test atmosphere. The test atmosphere contains a substance that typically can be detected by the wearer such as isoamyl acetate (banana oil), irritant smoke, or saccharin. The respirator must be equipped with the proper cartridges to remove the substance. For example, if using isoamyl acetate, which is an organic chemical which gives off a vapor, an organic-vapor chemical cartridge must be used. If the substance is irritant smoke, which is a particulate, a high-efficiency particulate air filter or HEPA cartridge must be used. If the wearer smells the isoamyl acetate or coughs due to the irritant smoke, the respirator is rejected and the test should be repeated until a respirator is found that fits properly. Figure 7.2 shows a trainee undergoing qualitative fit-testing.

These tests are relatively fast, are easily performed, and use inexpensive equipment. Because these tests are based on the subjective response of the wearer to the test chemical, the reproducibility and accuracy may vary.

Quantitative Fit-Testing Quantitative fit-tests involve exposing the respirator wearer to a test atmosphere containing an easily detectable, non-

192 ■ RESPIRATORY PROTECTION

FIGURE 7.2 ■ Worker undergoing qualitative fit-testing with irritant smoke.

toxic aerosol, vapor, or gas as the test agent. Instrumentation, which samples the test atmosphere and the air inside the facepiece of the respirator, is used to measure quantitatively the leakage into the respirator. With this information, a quantitative fit factor can be calculated. This factor is an index that indicates how well the respirator fits the wearer. The higher the number, the better the fit (see Table 7.1).

Table 7.2 was adapted in part from *3M Respirator Selection Guide and the ACGIH-TLVs*.

7.3 RESPIRATORY EQUIPMENT

Respiratory devices vary in design, application, and protective capability. After assessing the inhalation hazard, the user must understand the specific use and limitations of the available equipment in order to ensure proper selection. Respiratory protective devices are tested and approved by the National Institute for Occupational Safety and Health-Mine Safety and Health Administration (NIOSH-MSHA) for protection against a wide variety of in-

TABLE 7.2 ■ Selection of Proper Air-Purifying Respirator Cartridges

Chemical Name	IDLH	Remarks	Cartridge/Canister Type	ACGIH TLVs (ppm or mg/m³) TWA	STEL
Acetone	20,000	Narcotic	Organic vapor	750	1000
Ammonia	500	Extremely irritating	Ammonia	25	35
Benzene	1,000	Skin penetrant carcinogen	Organic vapor	(10, A2)[a] intended change to (0.3, A1)[a]	—
Dieldrin		Octalox—insecticide, skin penetrant	Organic vapor with high-efficiency filter combination	0.25 mg/m³	—
Silicon tetrachloride			High-efficiency	—	—
Tetrachloroethylene	500	Perchloroethylene, skin penetrant	Organic vapor	(25, A3)[a]	(100)
Tetraethyl lead	40[d]	As lead, skin penetrant	Organic vapor	0.1 mg/m³	—

[a]A1, confirmed human carcinogen; A2, suspected human carcinogen; A3, animal carcinogen.
[d]mg/m³
Source: Adapted, in part, from *3M Respirator Guide* and the ACGIH-TLVs.

halation hazards. Whenever a respiratory protective device is used, it is essential that the device be NIOSH-MSHA-approved. There are two basic types of respirators: air-purifying and atmosphere-supplying.

7.3.1 Air-Purifying Respirators (APRs)

Air-purifying devices remove contaminants from the atmosphere and are to be used only in atmospheres containing at least 19.5% oxygen by volume. The common types of air-purifying respirators are mechanical filter respirators, chemical cartridge respirators, combination of mechanical and chemical respirators, and gas masks. Powered APRs are also available. These units consist of a mask, battery, HEPA filter, fan, and motor.

Mechanical filter respirators provide protection against airborne particulate matter including dusts, metal fumes, mists, and smokes. They consist essentially of a soft facepiece of either half-mask as shown in Figure 7.3 or full-facepiece design as shown in Figures 7.4 and 7.5. Depending on the design of the mask, one or two mechanical filter elements which contain fibrous material (usually resin-impregnated wool) can be attached. The filter removes harmful particles by physically trapping them as air passes through it during inhalation; gaseous material passes through the filter. Figure 7.6 shows the air-flow patterns into and from an APR respirator. High-efficiency particulate air (HEPA) filters are used for dusts, fumes, and mists with TLVs less than 0.05 mg/m^3. These filters have very small pores and can be used for radionuclide filtration in the range of 0.3 μm.

Chemical cartridge respirators afford protection against light concentrations (10–1000 ppm by volume) of certain acid gases, alkaline gases, organic vapors, and mercury vapors by using various chemical filters to purify the inhaled air. In contrast to mechanical filters, the chemical cartridge respirators contain sorbents to remove harmful gases and vapors. These sorbents work on either an absorption or an adsorption principle and usually are either charcoal or silica gel. Because there is only a specified amount of sorbent in a cartridge, high concentrations of vapors will quickly saturate the material and cause a breakthrough. Concentrations for which individual cartridges are effective are provided by the manufacturer. Check for the NIOSH-MSHA approval before purchasing and using any cartridge.

Combination cartridges—that is, dust, mist, fume, or high-efficiency filters in combination with chemical filters—are used when field conditions or response actions indicate that dual or multiple hazards are present or could be present in the atmosphere.

7.3 RESPIRATORY EQUIPMENT ■ **195**

FIGURE 7.3 ■ Worker wearing a half-mask APR. *Source:* Courtesy of Scott Aviation.

Limits on APR Use Air-purifying respirators are effective as protective equipment when used within the limits of time and concentration for which they were designed. The following points must be addressed before selecting an APR and cartridge type:

- Because APRs do not supply air, they can only be used in oxygen-sufficient atmospheres, those with greater than 19.5% oxygen.
- IDLH concentration of contaminant must not be exceeded.

FIGURE 7.4 ■ **Full-facepiece APR.** *Source:* Courtesy of Scott Aviation.

- Odor threshold must be below TLV levels (gas or vapor must have good warning properties—that is, must be capable of odor detection below TLV level).
- Read the American National Standards Institute's (ANSI's) maximum use concentration recommendations before use of APR. Then review the following:

    1. Check manufacturer's listing of suggested cartridge to use.

FIGURE 7.5 ■ Trainee dressed in Level C protection with a full-facepiece APR.

2. Refer to odor threshold test data to ensure that odor threshold is well below TLV level; that is, the chemical should have good warning properties.
3. Maximum use concentration (MUC) for all organic vapor cartridges is 1000 ppm.
4. MUC for organic vapor gas mask canister is 20,000 ppm (chin style is 5000 ppm). It is suggested to use the 5000-ppm level for the gas mask canisters unless manufacturer recommends lower value.
5. Recommended MUC for acid gas cartridges are as follows unless the manufacturer recommends lower values:

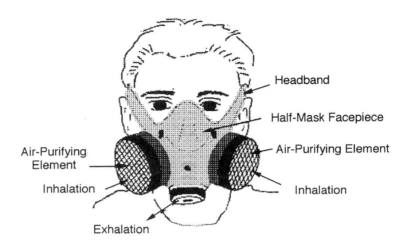

FIGURE 7.6 ■ Half-mask air-purifying respirator showing airflow patterns into and from the mask.

Hydrogen chloride (HCl) = 50 ppm
Sulfur dioxide (SO_2) = 50 ppm
Chlorine (Cl_2) = 10 ppm

7.3.2 Supplied-Air Respirators (SARs)

An air- or atmosphere-supplying respirator provides the highest level of protection possible for a site worker or response individual. It isolates the inidivdual's respiratory tract from any contaminated atmosphere. Supplied-air respirators fall into three main types:

1. Airline respirator
2. Combination airline respirator with egress bottle
3. Self-contained breathing apparatus (SCBA)

Airline Respirators These respirators are connected to a suitable compressed air source by a small-diameter hose. The air is supplied continuously under pressure in sufficient volume to meet the wearer's demand. Accessory equipment such as pressure regulators, pressure-relief valves, and air filters may be necessary to make sure the air is at the proper pressure and quality for breathing.

All airline respirators must be supplied with respirable air conforming to Grade D Compressed Gas Association's Standard CGA G-7.1-73, Commodity Specification for Air, 1973. This standard requires air to have the following content:

- Oxygen content normally present in the atmosphere
- No more than 5 mg/m^3 of condensed hydrocarbon contamination
- No more than 20 ppm of carbon monoxide
- No pronounced odor
- A maximum of 1000 ppm carbon dioxide

These airline respirators must be used only in non-IDLH atmospheres. Therefore, this type of airline respirator is not approved for emergency response or hazardous waste site operations. This limitation is necessary because the airline respirator is entirely dependent upon an air supply that is not carried by the wearer and that might fail, leaving the wearer without a safe atmosphere.

Combination Airline Respirators with Egress Bottle This type of supplied-air respirator is very similar to the simple airline respirators except that they have an egress bottle which contains approximately five minutes supply of air for escape as shown in Figures 7.7 and 7.8. If the air supply to the respirator fails for some reason such as compressor failure, hose blockage, and so on, the wearer can unplug the hose line and turn the bottle supply on, thereby giving him or her a five-minute supply of air for their escape. These respirators (with pressure-demand flow; see below) are approved for IDLH atmospheres, emergency response, and hazardous waste site operations.

Air-Flow Options There are three main types of airline respirators: constant-flow, demand-flow, and pressure-demand-flow. The respirators are equipped with half-masks, full facepiece, and helmets or hoods. These include:

Constant-Flow. A constant or continuous-flow unit has a regulated amount of air constantly fed to the facepiece and is normally used where there is an ample supply of air such as that supplied by an air compressor. These units are not approved for emergency response or hazardous waste site activities.

Demand-Flow. These airline respirators with half masks or full facepieces deliver airflow only during inhalation. A suitable regulator is

FIGURE 7.7 ■ Worker demonstrating an airline respirator with egress bottle. *Source:* Courtesy of Scott Aviation.

FIGURE 7.8 ■ Trainee dressed in Level B protection with airline respirator with egress bottle. Airlines are visible.

required to make sure the air is reduced to the proper pressure for breathing. Since negative pressure is created in the facepiece as a result of inhalation, the facepiece must fit tightly around the face while providing a good fit; otherwise contaminated air will be drawn in. Therefore, these respirators are not approved for emergnecy response or hazardous waste site operations.

Pressure-Demand. To solve the negative-pressure problem presented in the demand mode, manufacturers developed the pressure-demand airline respirator. The pressure-demand respirator provides positive pressure in the facepice during both inhalation and exhalation thereby preventing the development of negative pressure in the facepiece and the possibility of contaminant entry. These respirators when equipped with egress bottle are approved for emergency response and hazardous waste site operations.

The NIOSH-MSHA approval for all airline respirators requires that the following specifications are met:

- The maximum hose length must be 300 feet or 91.5 meters.
- The maximum permissible inlet pressure must be 125 psig or 683 kPa.
- The pressure to any length of hose assembled to the respirator must be sufficient to deliver 115 liters/min (4 cfm) measured at the facepiece.
- When helmets or hoods are used, the same requirements must be met, except that the flow rate must be at least 170 liters/min (6 cfm).

Self-Contained Breathing Apparatus (SCBA) The SCBA provides complete respiratory protection against toxic gases and oxygen deficiency. The wearer is independent of the surrounding atmosphere because the wearer is breathing with a system that is portable and admits no outside air. The units consist of a carrying assembly and bottle, tank or cylinder, pressure regulator, a gauge, a safety valve, and a full facepiece. The tank is equipped with an alarm to warn the wearer when air in the tank is getting low. Figures 7.9 and 7.10 show workers equipped with SCBA, and Figure 7.11 shows a worker with an SCBA equipped with a voice microphone.

Most SCBAs operate in an open-circuit mode. With an open-circuit SCBA, the exhaled air is vented to the atmosphere and not rebreathed. Air is supplied from the tank. These open-circuit units are available in the continuous-flow demand and pressure-demand modes. However, as with the airline respirators, only the pressure-demand mode is approved for emergency response and hazardous waste site operations.

Self-contained breathing apparatus are also available as oxygen cylinder rebreathing and self-generating types.

Oxygen Cylinder Rebreathing This unit has a relatively small cylinder of compressed oxygen, reducing and regulating valves, a breathing bag, facepiece, and chemical container to remove carbon dioxide from the exhaled air. The types of units currently manufactured are approved by NIOSH-MSHA for emergency response and hazardous waste site operations. These

FIGURE 7.9 ▪ Worker dressed in Level B protection with SCBA.
Source: Courtesy of Scott Aviation.

FIGURE 7.10 ▪ Trainee performing simulated exercises in Level B with SCBA.

units are available for 45-minute, 60-minute, 2-hour, 3-hour, or 4-hour durations. The following description covers all five types.

The high pressure of the oxygen from the cylinder is lessened by means of a reducing and regulating valve. Exhaled breath passes through a tube into a container holding the carbon dioxide-removing chemical and then through a cooler. Finally, the purified air flows into the breathing bag, where it mixes with the incoming oxygen from the cylinder.

The rebreathing principle permits most efficient use of the oxygen supply. The exhaled breath contains both oxygen and carbon dioxide as the human body extracts only a small part of the oxygen inhaled. As the user exhales, the carbon dioxide is removed by the chemical and the oxygen that

FIGURE 7.11 ▪ Worker demonstrating SCBA equipped with microphone for communication. Courtesy of Scott Aviation.

is left is used. This method of operation applies to all oxygen cylinder rebreathing-type apparatus as well as those using liquid oxygen.

The oxygen cylinder must be refilled and the carbon dioxide removing chemical replaced after each use. As is true of all respiratory protective equipment, training in proper use and maintenance is essential for the most efficient operation.

The Self-Generation Type This apparatus differs from the conventional cylinder rebreathing type in that it has a chemical canister that evolves oxygen and removes the exhaled carbon dioxide in accordance with breathing requirements. It eliminates high-pressure cylinders, regulating valves, and other mechanical components.

The canister, which contains potassium superoxide, evolves oxygen when contacted with moisture and carbon dioxide in the exhaled air and retains carbon dioxide and moisture. Retaining moisture is important because it aids in preventing lens fogging.

In use, the self-generating unit operates as other rebreathing apparatus except the wearers, using the canister, make their own oxygen instead of drawing it from a compressed gas cylinder or liquid oxygen source. The important features of this type are (a) the simplicity of construction and use and (b) reduced need for maintenance when compared with the high-pressure apparatus.

Use and Service of SCBA SCBAs differ by manufacturer and type. You should familiarize yourself with the manufacturer's instructions and checkout procedure before using any SCBA.

Because most emergency responders and site workers use the open-circuit, pressure-demand SCBA, the remaining discussion will focus on this type.

The key parts of an SCBA other than the air tank, facepiece, and hose include the following:

- The *main-line valve* (yellow) controls flow to the regulator. This valve should be left open except when using the bypass valve.
- The *bypass valve* (red) bypasses the regulator in case of malfunction of the regulator or of the main-line valve. This bypass should be open only when needed. The regulator should not be bled with this valve.

Placing an SCBA in Service

1. Follow manufacturer's instructions
2. Perform an operational inspection
3. Preparation for use
4. Use of respirator
5. Termination of use
6. Cleanup and maintenance

Donning an SCBA To put on an SCBA the following things must be performed (check the manufacturer's instructions for specific instructions).

1. Check gauges and valves.
2. Turn on cylinder valve and listen for low-pressure alarm.
3. Put on the tank and harness and adjust straps.
4. Don the facepiece and check the facepiece seal (fit-testing; see previous section.)
5. Check the main-line and the bypass valves.

The equipment should be donned according to the manufacturer's recommended procedures. Workers should be trained for each type of SCBA that will be used. Routine training and practice is necessary especially, for emergency responders who may use this equipment infrequently.

General Limitations of Open-Circuit SCBAs

- The period over which the device will provide protection is limited by the amount of air or oxygen in the tank, the ambient atmospheric pressure, and the type of work being performed.
- Main limitations are their weight, bulk, or both, limited service life, and the training required for their maintenance and safe use.

7.4 CARE OF RESPIRATORS

Like any piece of equipment, respirators require routine inspection, cleaning, and maintenance in order to ensure proper protection (see Figure 7.12). This inspection is especially important for equipment used infrequently (for emergency response). Without routine care, the equipment may become damaged, outdated, or misplaced.

208 ■ RESPIRATORY PROTECTION

FIGURE 7.12 ■ **Shows a facility used for the cleaning and testing of respirators.** *Source:* Courtesy of OHM Remediation Services Corp.

Some general guidelines on respirator care are given below. Specific requirements for the care of any particular respirator are listed in the manufacturer's literature.

7.4.1 Care of APRs

Inspect APRs before and after each use and check at least monthly even if the respirator has not been in use.

- Check the point where the cartridges screw into the mask, the valves and gaskets.
- Check the condition of the facepiece. If cracks or hardening of the facepiece material is noted, replace.
- Check the headbands to make sure that they can be tightened to provide a good fit.
- Check the harness for cracks or tears.
- Defects or unusual conditions should be reported immediately.

Cleaning and Disinfecting an APR
- Take the respirator apart.
- Wash everything except the particulate filters and the cartridges with disinfectant soap and water.
- Air dry.
- Reinspect each part as it is put back together.

Storing APRs
- Store away form dust, sunlight, heat, extreme cold, high humidity, and chemicals that could damage the APR.
- Store each respirator in a separate bag if possible.
- Follow manufacturer's instructions for specific storage requirements.

7.4.2 Care of SCBAs

Inspect SCBAs before and after each use and check at least monthly even if the respirator has not been in use. A company policy may include more frequent inspections to ensure that SCBAs are ready for use.

- Check the condition of the facepiece; replace if cracks or hardening is noted.
- Check the hose and the points where the hose attaches to the facepiece and to the air tank.
- Check the headbands to make sure that they can be tightened to provide a good fit.
- Check the head and tank harnesses for cracks, tears, or other defects.
- Check the regulator according to the manufacturer's directions.
- Check the air tanks for damage.
- Defects or unusual conditions should be reported immediately.

Cleaning and Disinfecting a SCBA
- Remove the air tank.
- Inspect each piece of the SCBA.
- Wash the facepiece, hose, and harness with disinfectant soap and water.
- Air dry.
- Reinspect each piece of the SCBA as it is put back together.

Storing SCBAs

- Store away from dust, sunlight, heat, extreme cold, high humidity, and chemicals that could damage any part of the SCBA.
- Tanks should be carefully stored to prevent mechanical damage.
- Follow manufacturer's instructions for storing specific types of SCBAs.
- Do not submerge SCBAs in water.

Note: Communication devices should be cared for according to manufacturer's recommendations.

7.4.3 Care of Airline Respirators

Care is similar to SCBAs, with the exception of tank and support harness care.

7.4.4 Respirator Program

OSHA requires that employers, who make respirators available to their employees, have a written respirator program (29 CFR 1910.134). The program should be evaluated at least annually (or when requirements change) and modified to reflect changes in the work place.

The respirator program should be included in or referenced in the Emergency Response Plan or the Site Safety Plan. Special considerations that may be included are:

- Need for corrective lenses in full facepiece respirators
- Restriction on use of contacts
- Communication needs
- Use in dangerous atmosphere, including confined spaces
- Use in extreme temperatures

Persons using respirators under the above conditions, or under other unusual conditions, should review special requirements with supervisors or health and safety representatives.

The respirator program should include a description of who is responsible for the various aspects of the program including selection, periodic and routine fit-testing, inspection, cleaning, repair, and maintenance.

Section 1910.134(e)(5)(i) of the OSHA General Industry Standards states:

Respirators *shall not be worn* when conditions *prevent a good face seal.* Such conditions may be *growth of beard*, sideburns, a skull cap that projects under the facepiece, or temple pieces on glasses... . To assure proper protection, the facepiece fit shall be checked by the wearer each time he puts on the respirator. [emphasis added]

OSHA's interpretation of this section is that there cannot be any facial hair when using any respirator which relies upon a good face-to-facepiece seal and which includes any tight-fitting (as opposed to helmet or loose-fitting hood) air-purifying respirator. Even several days' beard growth or a heavy stubble can reduce the possibility of a face-to-facepiece seal.

The question often comes up, "Can an employer force a worker to shave their beard because they will have to wear a respirator?" The answer depends in part on whether the use of a respirator was a condition of employment. If the employee was hired with the understanding that the use of a respirator was part of the job responsibilities, then the employer can expect the employee to be clean-shaven.

The manufacturers of respirators attach a warrantee to their facepieces that states that these respirators cannot be used on persons with facial hair, protruding dentures, or facial scars that prevent a proper face–facepiece seal.

7.5 RESPIRATORY PROTECTION PROGRAM

The generic respiratory protection program below was provided by the Occupational Safety and Health Administration (OSHA).

RESPIRATORY PROTECTION PROGRAM

Proposed for _____

This respiratory protection program is established to coordinate the use and maintenance of respiratory protection equipment which is used to reduce employee exposure to air contaminants. In addition, it will allow employees to work safely in potentially hazardous environments.

Established _____
                              (*date*)

_____
      (*executive officer*)

The administration of the overall respiratory protection program will be the responsibility of _____

Administrative responsibilities include:

- A. Identification and location of hazardous exposures,
- B. Supervision of respirator selection.
- C. Supervision of medical screening for potential respirator users.
- D. Supervision of employee training and qualitative respirator fit testing.
- E. Supervision of cleaning, maintenance and storage of respirators.
- F. Evaluation of overall respirator program.

A. Identification and Location of Air Contaminant Exposures

Based on a comprehensive industrial hygiene evaluation conducted on _____, all potentially hazardous air contaminant exposures are summarized in A-1. Additional air contaminant monitoring will be conducted whenever exposures are expected to change. For example, whenever new raw materials are used, production processes change or a spill occurs.

This monitoring will be conducted by _____. Subsequent information will be added to A-1 as it is accumulated.

B. Respirator Selection

All respirators will be selected based on the criteria spelled out by ANSI Z88.2-1980 which is summarized in B-1. B-2 serves as an information guide for the selection process.

B-3 is a summary of the respirators presently in use. This summary will serve as a guide for ordering new respirators and replacement parts. Orders will be initiated by _____ through _____.

Changes to this list will be updated as necessary.

C. Medical Evaluation for Users of Respirators

Prior to assignment to any position requiring respirator use, a medical evaluation of the employee's physical ability to work while wearing a respirator will be necessary. An outline of the evaluation is found in C-1. A periodic evaluation identical to the preplacement evaluation will be done every 2 years. If a change in the employee's medical condition occurs, the interval for periodic evaluation will be established by a physician.

C-1 and the respirator to be worn will be sent along with the employee for the examination. Physicians approval in C-1 will be necessary before a respirator will be assigned.

D. Employee Training and Fit Testing

Training in the use and limitations of respirators will be provided to all respirator users. Initial training along with refresher training will be conducted by _____. D-1 will serve as a guide for the training as well as a documentation of training dates. During the training employees will be advised of the potential hazards associated with excessive exposure as summarized in the health guidelines in D-2.

Qualitative fit testing will be performed by _____ as part of the employee trainee training program and annually thereafter. A record of the tests will be maintained in D-3.

E. Respirator Cleaning, Maintenance and Storage

Cleaning and maintenance of respirators will be the responsibility of _____. Respirators will be individually issued and a record of issuance will be maintained in E-1. Information on:

1. Receipt date
2. Location
3. Monthly inspection records for both emergency and extra routine use respirators.

Procedures for cleaning, maintenance and storage are outlined in E-2.

F. Respirator Program Evaluation

The overall evaluation of the respirator program will be conducted by _____ on a monthly basis. This evaluation will include inspection of records contained in D-1 and E-1, observation of user proficiency, and random inspection of respirators for cleanliness, deterioration, proper selection and proper storage. A record of the evaluation will be recorded in F-1.

A-1. Identification and Location of Air Contaminant Exposures

Location	*Operation*	*Contaminants*	*Exposure*	*Date*

B-1. Respirator Selection Criteria
Selection Criteria

1. Based on the minimum protection factor needed, select the respirator type from ANSI Z-88, 2-1980, Table 5.

2. Review specific contaminant information in B-2.

3. If skin irritation or absorption is a problem, a respirator alone will not provide complete protection and special protective clothing will be needed.

4. At what level of exposure are there noticeable warning properties which can signal respirator failure? For example, at what level is the odor noticeable and

how does this relate to the workers' actual exposure. If warning properties are nil, then supplied air is the only alternative.

5. Is eye irritation a problem at the exposed concentrations? If so, a full facepiece will be needed.

6. If the concentrations approach the level of the IDLH or the LEL, then the only acceptable respirator is SCBA or supplied air with egress bottle (auxiliary SCBA).

7. If the contaminant is a particulate, consider the health effects. If it is a systemic poison, then no single use respirator is acceptable.

8. If it is a particulate with a PEL of less than 50 micrograms/m^3, then a high efficiency filter is needed.

9. Where gases and vapors are of concern, is the sorbent efficiency of the air purifying canister or cartridge satisfactory? If not, supplied air is necessary.

10. Select only respirators and cartridges having NIOSH/MSHA approval.

B-2. Selection Information

General Information
1. Work Description/Operation:
2. Anticipated Use Time:
3. Worker Activity Level:
4. Work Area Location:
5. Work Area Characterization:
6. Location of Hazardous Area Relative to Safe Area:

Specific Information
1. Oxygen Content: _____%
2. Air Contaminants:
   Chemical name
   Trade name
   Physical state
   (dust, fume, mist, gas, vapor)

3. Exposure Limits:
   OSHA 8-hr TWA
   OSHA Ceiling
   ACGIH 8-hr TWA
   ACGIH Ceiling

RESPIRATORY PROTECTION

NIOSH 8-hr TWA _____ _____ _____
NIOSH Ceiling _____ _____ _____
Other   _____ _____ _____
               _____ _____ _____

4. Warning Properties:
   Eye irritation
   concentration _____ _____ _____
   Respiratory irritation
   concentration _____ _____ _____
   Odor threshold
   concentration _____ _____ _____

5. IDLH Concentration: _____ _____ _____

6. Skin Absorption
   (yes or no) _____ _____ _____

7. Skin Irritation
   (yes or no): _____ _____ _____

8. Chemical Properties:
   Vapor pressure _____ _____ _____
   Lower flammable
   limit (LEL/LFL) _____ _____ _____
   Upper flammable
   limit (UEL/UFL) _____ _____ _____

9. Sorbent Efficiency: _____ _____ _____

10. Minimum Respirator
    Protection Factor
    Needed _____ _____ _____

B-3. Respirator Selection Summary

Location	Operation	Respirator	Use	Air Contaminants	NIOSH	Approval

C. Medical Evaluation for Users of Respiratory Protective Equipment

C-1. Medical Evaluation Outline

Employee: _____ Respirator: _____

Date: _____ Job Description: _____

Estimated Respirator Use Time: _____

Work Activity: _____

Air Contaminants Exposed to: _____

1. A medical history emphasizing the presence and degree of cardiopulmonary complaints, e.g., dyspnea, cough, sputum production, wheezing, and exertional chest pain. A smoking history should also be elicited.

2. A comprehensive occupational history detailing prior respirator use, if any, and any difficulties experienced.

3. A physical exam with special emphasis on the cardiopulmonary systems.

4. A 14- by 17-inch posterior–anterior chest roentgenogram (X-ray).
5. Pulmonary function studies including forced respiratory volume in one second (FEV_1) and forced vital capacity (FVC).
6. A resting electrocardiogram for those workers over 40 years of age or any who exhibit any signs of symptoms of cardiac disease.
7. Individuals with any of the following conditions may not be qualified for respiratory use:
    a. Symptomatic pulmonary disease with moderate-to-severe ventilatory impairment, e.g., chronic bronchitis, emphysema, asthma.
    b. Symptomatic, severe or progressive cardovascular disease, e.g., uncontrolled hypertension, angina pectoris.
    c. Any anthropometric features which preclude an adequate seal, e.g., deep facial scars, blemishes, excessive facial hair, hollow temples, receding chin.
    d. Any psychological condition where the employee cannot tolerate respirator use, e.g., claustrophobia, severe anxiety.
    e. Any skin hypersensitivity to the chemical components of the respirator itself.
    f. Severe arthritis; absence of fingers, hand, or arms; or any other condition interfering with manual dexterity necessary for proper respirator use.
    g. Any required vision correction incompatible with emergency respirator use.
    h. Impaired sense of smell which may reduce warning properties of contaminant.
8. Individuals with a mild degree of pulmonary impairment should be further evaluated to determine their ability to wear a respirator. A trial period with the individual wearing a respirator in a nonhazardous atmosphere is suggested.

Medical Report Request Form

Dear Dr. _____:

It is our company policy that before a worker can be required to wear respiratory protection on the job, a medical examination is needed to determine if the worker is capable of wearing the protective device.

On the preceding page is some information pertaining to the type of work performed, the respirator to be used, and a recommendation of what the examination should include.

Upon completion of the examination, please complete the following and return to me.

Based on my evaluation, Mr./Ms. _____

    Has no medical condition which would be aggravated or interfere with the use of respiratory protection.

    Should not be required to wear respiratory protection.

Doctor's signature _____

Date _____

Thank you,

_____

D. Employee Training and Fit Testing

D-1. Respirator Training

1. The user will be instructed in the nature of the hazard or hazards for which respiratory protection is being provided and informed of the possible consequences which may occur if exposed to the hazard without adequate protection. Health hazard guidelines are contained (supplied by the company). The user will also be made aware that every reasonable effort is being made to reduce or eliminate the hazard.

2. Instruction will include a discussion of the respirator's capabilities and limitations along with the functional parts of the device and the possible malfunctions of the each part.

3. A detailed discussion of the user's responsibility for the inspection of the equipment prior to use and the appropriate points of inspection will be included. Each user will have access to the respirators we will be required to use during this part of the training.

4. Instruction and training will include guidance on proper storage, method of obtaining cleaning and maintenance service, and methods to assure adequate fit and function of the device each time it is donned.

5. Instructions on obtaining equipment, donning methods, proper fitting, and adjustment of the equipment will be given. Each user will then don the equipment in an atmosphere of normal air, prior to a fit-testing exercise.
6. Qualitative fit-testing.
7. A record of employee names, dates, type of initial training, and subsequent refresher training will be maintained.

TRAINING RECORD

Name	*Department*	*Respirator Type*	*Use*	*Date*

Initial all dates.   Signature _____

Respirator Fitting Test

(Qualitative fit-testing)

Name: _____ Date: _____

Job: _____ Glasses Worn: _____

Facial hair, dentures, etc. _____

_____

RESPIRATOR TYPE

_____

_____

7.5 RESPIRATORY PROTECTION PROGRAM ■ 221

A. Compatible with eye glasses

_____

_____

B. Irritant smoke test
  (1) Head stationary, normal breathing
  (2) Head stationary, deep breathing
  (3) Head turning side to side
  (4) Head moving up and down
  (5) Talking

_____

_____

C. Comfort
  (1) Very comfortable
  (2) Comfortable
  (3) Barely comfortable
  (4) Uncomfortable
  (5) Intolerable

_____

_____

Comments:

Tested by:
Signed _____ Date _____

E. Respiratory Program

E-1. Respirator Use and Maintenance Record

  Respirator Type

■ RESPIRATORY PROTECTION

Manufacturer

Model Number

    Date placed in service _____
    Assigned to whom   _____

Inspection and Maintenance Record

Date   Inspected/Serviced By

_____

_____

_____

_____

_____

_____

_____

_____

_____

_____

_____

_____

_____

_____

E-2. Respirator Cleaning, Maintenance, and Storage

The respirator *must* be washed and disinfected after each day of use by carrying out the following instruction:

1. If using an air-purifying respirator, remove the air-purifying element from each respirator. An air-purifying element must never be washed and disinfected.
2. Immerse the respirator in a warm (140–160°F) water solution of a mild detergent. The respirator facepiece and parts may be scrubbed gently with a cloth or soft brush. Make sure that all foreign matter is removed from all surfaces of the rubber exhalation valve flap and plastic exhalation valve seal. Repeat the cleaning procedure for the inhalation valve and seal, and the inner and outer surfaces of the respirator.
3. After washing, rinse in clean warm water and shake excess water from mask.
4. Immerse the respirator in a mild chlorox (or comparable disinfectant) solution (approximately 1/4 cup of 3% chlorox/gallon warm water). Allow respirator to sit in disinfectant (140–160°F) for 10–20 minutes.
5. After washing and disinfecting the respirator, rinse the respirator in several changes of clean, warm water and then allow the respirator to air dry (hair dryers can be used but only with the cold air setting).
6. When required for use, select and attach a new air-purifying element.

STORAGE

When the respirator is not in use, it should be placed in a plastic bag or plastic box and then stored in designated cabinet, locker, or chest. Respirators should stored in a cool dark place in a single layer with the facepiece and exhalation valve in a more or less normal position to prevent the facepiece material from taking a permanent distorted shape.

F. Respirator Program Evaluation

1. Are all records complete and up to date? Yes _____ No _____
   If no, what action has been taken to improve future performance?
   _____
   _____
2. Has air contaminant monitoring been conducted at operations where new materials or activities are being used and/or performed? Yes _____ No _____
   If no, what action has been taken to determine exposure?

3. Are employees wearing the respirators which have been selected for the job?
   Yes _____ No _____
   If no, what action has been taken to eliminate the use of improper respirators?

4. Do all employees wearing respirators have medical approval and fit test?
   Yes _____ No _____
   If no, what is being doine to correct the situation?

5. Have all employees completed their initial or refresher respirator training?
   Yes _____ No _____
   If no, what is being done to complete training?

6. Do employees who have completed training understand limitations, use, maintenance and inspection of respirations? Yes _____ No _____
   If no, what improvements in the training program are being implemented?

Date _____   Signature _____

REVIEW QUESTIONS

1. List several situations in which respiratory protection would be required.
2. What are the limitations of APRs?
3. What are the limitations of SCBAs?
4. Why are routine positive- and negative-pressure tests important?
5. Why are medical exams required for persons who use respirators?
6. List parts of a respirator which should be checked before and after each use.
7. Why is proper storage of respirators important?
8. List items that must be included in a written respiratory protection plan.

REFERENCES

American National Standards Institute (ANSI), *Standard Z88.2-1980*, ANSI, New York, 1980.

Levine, S., and W. Martin,: eds., *Protecting Personnel at Hazardous Waste Sites*, Butterworth, Stoneham, MA, 1985.

3M-Occupational Health and Environmental Safety Division, *3M™ Respirator Selection Guide*, 3M, St Paul, MN, 1995.

Mine Safety Appliance (MSA), *Response Respirator Selector*, MSA, Pittsburgh, PA, 1995.

NIOSH/OSHA/USCG/EPA, *Safety and Health Guidance Manual for Hazardous Waste Site Activities*, U.S. Government Printing Office, Washington, D.C., 1985.

OSHA. 29 CFR 1910.134, *Respiratory Protection*, U.S. Government Printing Office, Washington, D.C., 1986.

MONITORING

CHAPTER OBJECTIVES

When you have completed this chapter, you will be better able to

- Understand why the work environment is monitored
- Identify hazards that can be monitored
- Understand what the air monitoring results mean and how to use the information
- Introduce several types of air monitoring instruments:
  Direct-reading equipment
  Personal sampling instruments

8.1 INTRODUCTION

Monitoring is a continuing program of observation, measurement, and judgment—all of which are necessary to recognize potential health hazards and to judge the adequacy of protection. Monitoring requires an awareness of the presence of potential health hazards and an assessment on a continuing basis of the adequacy of the control measures in place.

Monitoring provides vital information about the hazards present at the site. Usually readings or samples with air monitors will be taken of the atmosphere surrounding the incident or site. Although no single instrument can detect all hazards, using air-monitoring equipment properly can provide information which is needed to protect life and property. However, using the wrong instrument or one that isn't working properly may endanger life and property by indicating conditions as not hazardous when, in fact, they are.

Monitoring will generally be conducted as part of risk assessment (size-up of the scene). Continuous monitoring helps the workers by continuously making them aware of the concentrations of chemicals and changes that may occur as a result of the site activities.

8.2 PURPOSE OF AIR MONITORING

Airborne contaminants can present a significant threat to human health, thus, identifying and quantifying these contaminants by air monitoring is an essential component of a health and safety program. Reliable measurements of airborne contaminants are necessary to

- Determine the level of personal protective equipment required
- Delineate areas where protection is needed
- Assess the potential health effects of exposure
- Determine whether continual exposure is occurring, which may indicate the need for medical monitoring.

8.2.1 Understanding Concentration Measurements

Concentration is the amount of substance contained in a certain volume of air. Concentration of gases and vapors are usually measured in parts per million (ppm):

- 1 part per million is equivalent to 1 inch in 16.7 miles.
- 100 ppm is equivalent to 1 teaspoon in 1300 gallons.

Concentrations of dusts and mists are usually measured in milligrams per cubic meter of air (mg/m^3):

- Approximately 400,000 milligrams are in one pound.
- Approximately 35 cubic feet are in one cubic meter (a meter is about 40 inches).

Concentrations of fibers are measured in fibers per cubic centimeter (f/cc):

- A cc is approximately the size of a sugar cube.

8.3 MEASURING INSTRUMENTS

The purpose of air monitoring is to identify and quantify airborne contaminants in order to determine the level of worker protection needed. Initial screening for identification is often qualitative, that is, the contaminants or the class of chemicals are demonstrated to be present but the determination of its concentration (quantification) must await subsequent testing. Two principal approaches are available for identifying and/or quantifying airborne contaminants:

- The on-site use of direct-reading instruments.
- Laboratory analysis of air samples obtained by "grab" sampling, i.e., the use of sampling bags, filters, sorbents, or wet-contaminant collection methods.

Characteristics of Air Monitoring Instruments To be useful in the field, air monitoring instruments must be

- Portable
- Able to generate reliable and useful results
- Selective and sensitive
- Inherently safe

All of these traits may not be present in any one instrument.

Portability Portability is a prime consideration for any field instrument. Transportation shock, together with unintentional abuse, greatly shortens the useful life of an instrument. To reduce this trauma, instruments should be selected that have reinforced shells or cases, shock-mounted electronic packages, or padded containers for shipment.

Equipment exposed to the elements and to test atmospheres is susceptible to breakdown. Anodized or coated finishes, weather-resistant packaging, and remote sensing are effective in reducing downtime and increasing portability.

In short, a portable unit should possess ease in mobility, the ability to withstand the rigors of use, quick assembly, and short check-out and calibration time.

Reliable, Useful Results Response time (the interval between the time the instrument senses a contaminant and when it generates data) is very important. Response times for direct-reading instruments range from a few seconds to several minutes. Another important factor is the operating range (difference between the lower detection limit or the lowest concentration the instrument will respond to and the saturation concentration or the upper use limit). Also important is the precision or reproducibility of the instrument (the extent to which the instrument will show the same or nearly the same reading under identical situations). The final necessary characteristic is accuracy or the ability of the instrument to accurately measure the concentration of the contaminant present.

Selectivity/Sensitivity Selectivity is the ability of an instrument to detect/measure a specific chemical or group of chemicals, while sensitivity is the ability of an instrument to accurately measure small concentrations of a particular contaminant.

Inherent Safety Electrical devices, including instruments, must be constructed in such a way as to prevent ignition of a combustible atmosphere. The source of such ignition could be an arc generated by the power source itself or the associated electronics, or a flame or heat source necessary for functioning of the instrument. The National Fire Protection Association (NFPA) has created minimal standards in the National Electrical Code for electrical devices used in flammable or ignitable atmospheres. The standard provides for three methods of construction to prevent a potential source from igniting a flammable atmosphere: explosion-proof, intrinsically safe, and purged. Make sure that all instruments used in hazardous environments meet NFPA standards.

Direct-Reading Instruments Direct-reading instruments were developed as early warning devices for use in industrial settings, where a leak or accident could release a high concentration of known chemical into the workplace. They were used to detect and/or measure high exposure levels that are likely to cause acute reactions.

Direct-reading instruments provide information at the time of monitoring and do not require sending samples to the lab for analysis. This characteristic of direct-reading instruments enables real-time decision-making.

These instruments may be used to rapidly detect conditions that are immediately dangerous to life and health (ILDH), or to demonstrate the presence of specific chemicals or classes of chemicals. They are the primary tools of initial site or response characterization. The information provided by direct-reading instruments can be used to institute appropriate protective measures (i.e., personal protective equipment, evacuation, engineering controls), to determine the most appropriate equipment for further monitoring, and to develop optimum sampling and analytical protocols.

All direct-reading instruments have inherent constraints in their ability to detect hazards:

- They can usually detect and/or measure only specific classes of chemicals.
- Generally, they are not designed to measure low or minute concentrations of contaminants.
- Many of the direct-reading instruments that have been designed to detect one particular substance also detect other substances (cross-sensitivity); consequently, these substances may cause interference and give false readings.

For all these reasons, it is imperative that these instruments be operated, and their data interpreted, by qualified individuals who are thoroughly familiar with the instrument's operating principles and limitations.

At hazardous waste sites, the presence of unknown and multiple contaminants are expected. Therefore, instrument readings should be interpreted conservatively. The following recommendations may facilitate accurate recording and interpretation:

- Calibrate instruments according to manufacturer's instructions before use.
- Develop chemical response curves if these are not provided by the manufacturer.
- Remember that the instrument's readings have limited value where contaminants are unknown. When recording readings of unknown

contaminants, report them as "needle deflections" or "positive instrument response" rather than specific concentrations (i.e., ppm). Conduct additional monitoring at any location where a positive response occurs.
- Report a reading of zero as "no instrument response" rather than "clean" because there may be quantities of chemicals present that are not detectable by the instrument.
- Repeat the survey with several detection systems to maximize the number of chemicals detected.

8.3.1 Combustible and Oxygen-Deficient Atmospheres

Combustible Gas Indicator (CGI)/Oxygen Meter Many combustible gas indicators (CGIs) and oxygen meters are constructed as two completely separate indicators in a single housing as shown in Figure 8.1.

Oxygen meters are used to evaluate an atmosphere for the following:

- Oxygen (O_2) content for respiratory purposes. The normal concentration of oxygen at sea level is 20.8%. Most oxygen meters are set to alarm if the oxygen falls below 19.5%.

FIGURE 8.1 ■ **Combustible gas indicator and oxygen meter model 261.** *Source:* Courtesy of MSA (Mine Safety Appliances Co.).

- Increase risk of combustion. Generally, concentrations above 23.5% are considered oxygen-enriched and increase the risk of combustion. Most meters will alarm when the meter goes above 23.5% O_2.
- Use of other instruments. Some instruments require sufficient oxygen for operation (some combustible gas meters do not give reliable results at oxygen concentrations below 10%).
- Presence of contaminants. A decrease in oxygen content can be due to the consumption (by combustion/hot work or a reaction such as rusting) of oxygen or the displacement of air by a gas or vapor of heavier density. If the oxygen decrease is due to consumption, then the concern is lack of oxygen. If it is due to displacement, there is something present that could be flammable or toxic.

Limitations in use of the oxygen meter include:

- Operation of oxygen meters depends on the absolute atmospheric pressure. Pressurized or low-pressure samples will give erroneous results. For atmospheric sampling at higher or lower altitudes the oxygen meter should be calibrated at the elevation where sampling is to take place.
- High concentrations of carbon dioxide (CO_2) shorten the useful life of the oxygen sensor.
- Temperature can affect the response of oxygen meters. Check the operations manual to ascertain the temperature compensation range for your instrument. Use of meter outside this range may be possible if meter is calibrated to temperature at which sampling will take place. At low temperatures the response of the meter is slowed, and below 0°F the sensor may be damaged by freezing.
- Strong oxidizing chemicals, like ozone and chlorine, can cause increased readings and indicate high or normal oxygen content when the actual content is normal or even below safe levels.

Combustible gas indicators measure the concentration of flammable vapor or gas in air. The results are displayed as the percentage of the lower explosive limit (LEL) of the calibration gas. Other flammable gases and vapors will cause a needle deflection or reading on the meter, but these results will be inaccurate. Always use the proper meter or calibration gas for the combustible gas you may encounter. Figure 8.2 shows a combustible gas meter with a hand pump. Because combustible gas indicators read the con-

234 ■ MONITORING

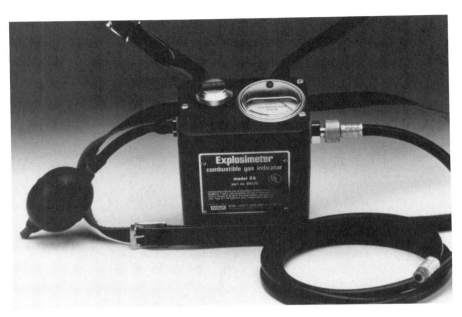

FIGURE 8.2 ■ **Combustible gas indicator with a manual hand pump.** *Source:* Courtesy of MSA (Mine Safety Appliances Co.).

centration of flammable gases as a percent of the LEL, it is important to understand what explosive limits are.

Explosive Limits
- The lower explosive limit (LEL) is the lowest concentration of a flammable substance in air which is required for ignition. Concentrations below the LEL will not ignite. Concentrations below the LEL are termed "too lean" to burn.
- The upper explosive limit (UEL) is the maximum concentration of a flammable substance in air which is required for ignition. Concentrations above the UEL will not ignite. Concentrations above the UEL are termed "too rich" to burn.
- The explosive (or flammable) range is the concentration of a substance in air between the LEL and UEL. In this range, the substance will readily ignite.

Reading A in Figure 8.3 shows meter reading of approximately 30% of the LEL. If this meter was calibrated for pentane, which has an LEL of 1.5%,

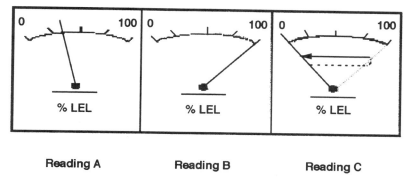

Reading A   Reading B   Reading C

FIGURE 8.3 ■ Three different meter readings or environmental conditions that could occur on site or during a response.

then the air would contain approximately 0.5% pentane. This is above the suggested safe levels for pentane and would warrant immediate evacuation.

Reading B shows a meter reading of 100% of the LEL or 1.5% pentane, in the explosive range, and a situation no worker should allowed to be in. Reading C shows another very serious condition. The concentration of combustible gas was so high that the meter went off scale and then returned to zero. In this situation the concentration could be above the UEL or still in the very high explosive range. This can occur if the concentration of pentane or other gas is too rich to burn, and therefore the meter reacts as if no combustibles were present. If the worker or responder was not paying attention, this meter deflection would have been missed and a serious condition could have occurred.

The most widely used CGI are calibrated for combustible vapors and gases such as pentane, hexane, and methane. It is important that you have at least a preliminary indication of the combustible gases you may encounter during your activities and choose a meter that is calibrated to those conditions.

The CGI will only detect combustible gases or vapors in air. It will not detect the presence in air of combustible airborne mists or dusts such as lubricating oils, coal dust, or grain dust.

The CGI uses a combustion chamber containing a heated filament that burns the flammable gas. The higher the concentration of flammable gas in the chamber, the higher the temperature of the filament (the higher the electrical resistance) and the higher the readings on the meter.

Limitations in the use of the CGI include:

- The CGI is intended for use only in normal-oxygen atmospheres. Oxygen-deficient atmospheres will give lower readings. Also, the safety guard that prevents the combustion source from igniting a flammable atmosphere is not designed to operate in an oxygen-enriched atmosphere.
- Certain materials in the sampled atmosphere affect the catalytic material on the filament and may cause the indicator to respond incorrectly. These materials include organic lead compounds such as those used in leaded gasoline and certain silicon compounds in the form of silanes, silicones, and silicates (often found in hydraulic fluids).
- Meter calibration and calibration gas is very important. Many CGIs are calibrated to pentane or hexane. The use of meters calibrated to these gases in situations involving other combustible gases will cause incorrect meter responses. These incorrect responses are called *relative responses*, and the manufacturer of the instrument may provide a table of relative responses for different gases (see Table 8.1).
- The relative response of the meter may be less than or greater than the meter's response to the same concentration of the calibrant gas (pentane or hexane). In other words, the meter may understate or overstate the actual concentration of gas being detected. If the identity of the gas is known, the actual concentration can be calculated; but this cannot be done if the identity of the detected gas is unknown.

Factors to consider when using a CGI include the following:

TABLE 8.1 ■ **Relative Response of Selected Chemicals for the MSA 260 Combustible Gas Indicator Calibrated to Pentane**

Chemical	LEL (%)	Concentration % LEL	Meter Response % LEL	Relative Response (%)
Methane	5.3	50	85	170
Acetylene	2.5	50	60	120
Pentane	1.5	50	53	106
1,4-Dioxane	2.0	50	37	74
Xylene	1.1	50	27	54

- Upon initially opening and/or probing an enclosed area, move the probe into the area slowly while watching the meter for an indication of a potentially dangerous condition. When evaluating confined spaces, sample the space at various levels to ensure that stratification of gases with differing vapor densities is not occurring.
- If you are using an extension tube on a CGI with a pump, make sure you allow enough time for the sample gas to displace the gas in the tube before you record or make note of the meter reading (see manual for time needed to clear ambient air per foot of tubing).
- If reading approaches or exceeds 10% of the LEL, extreme caution should be used in continuing the investigation. If reading approaches or exceeds 20% of the LEL, personnel should be withdrawn immediately. However, in view of the possibility of understated (or overstated) readings when unknown contaminants are detected, the project safety officer may modify these guidelines.
- An atmosphere that shows no flammability hazard can still be toxic to workers. Also, a confined space which is safe before work is begun may be rendered unsafe by work activities which cause a temperature increase, vaporization of sludge or scale, or a reduction in oxygen concentration.

8.3.2 Toxic Atmospheres

Detecting the presence and/or measuring the concentration of toxic chemicals is of utmost importance to the site worker and the first responder. However, the ability to determine the specific hazard by monitoring the air will be limited to the capabilities of the monitoring instrument used. Figures 8.4 and 8.5 show several toxic gas alarms for H_2S, O_2 and CO.

Toxic Gas/Vapor Detectors Toxic gas detectors contain specific sensors (hydrogen sulfide, carbon monoxide, hydrogen cyanide, chlorine, sulfur dioxide, etc.). These detectors can contain one or more sensors and are commercially available in combination with CGIs and oxygen meters. Figure 8.6 shows such a multigas detector.

Most toxic gas sensors and all oxygen monitoring instruments use electrochemical sensors. These sensors have a finite life, usually about one year.

Toxic gas electrochemical sensors need to maintain a bias in order to maintain a stable zero. Therefore, they require power to them to keep their bias. A toxic gas instrument is not off even though the instrument is in the off mode.

238 ■ MONITORING

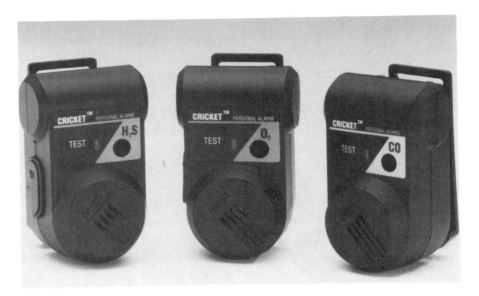

FIGURE 8.4 ■ Several different toxic gas alarms for H_2S, O_2, and CO. These indicators alarm at set value when the ambient air concentration reaches or exceeds the alarm concentration. *Source:* Courtesy of MSA (Mine Safety Appliances Co.).

Toxic gas electrochemical sensors are usually quite sensitive and typically can detect concentrations to 1 ppm. These sensors are fairly fast-reacting with reading possible in one minute or less. They are fairly specific, but there are some interference gases. Be familiar with these gases when using specific sensors.

The newer toxic gas detectors are self-calibrating and have digital readouts as shown in Figure 8.7.

Organic Vapors/Gases The direct-reading instruments to be discussed in this section for the purpose of monitoring toxic and/or flammable organic vapor containing atmospheres are

- Photoionization detectors
- Flame ionization detectors

Photoionization Detectors (PID) When the presence or types of organic vapors/gases are unknown, direct-reading instruments should be used. These instruments respond to all organic gas/vapors with ionization poten-

FIGURE 8.5 ■ Personal toxic gas alarms for H₂S, O₂, and CO. *Source*: Courtesy of Gas Tech, Inc.

tials (energy required to remove an electron) less than the ionization potential (energy output) of the lamp used in the detector.

Until the specific constituents can be identified, the readings from such instruments should be considered and recorded as total airborne substances to which the instrument is responding. Identification of the individual vapor/gas constituents may permit the instrument to be calibrated to these substances and used for more specific and accurate analysis.

Sufficient data can be obtained with the use of a PID during the initial entry to map or screen the site for various levels of organic vapors as shown in Figures 8.8 and 8.9. These gross measurements can be used on a preliminary basis to

- Determine the levels of personal protection
- Establish site work zones
- Select areas for more thorough study (i.e., sampling)

This instrument provides the user with the ability to assess the work environment for the presence of many organic gases/vapors. It employs the prin-

240 ■ MONITORING

FIGURE 8.6 ■ MSA Passport® five gas indicator with separate alarms. *Source*: Courtesy of MSA (Mine Safety Appliances Co.).

ciple of photoionization. This process involves the absorption of ultraviolet light (a photon) by a gas molecule, leading to its ionization. The sensor consists of a sealed ultraviolet light source that emits photons with an energy level high enough to ionize many organic compounds, but not high enough to ionize the major components of air. Ions formed by adsorption of photons are collected on an electrode, producing a current which is measured and converted to a concentration reading on the meter (in ppm).

The analyzer consists of a probe or sample inlet, a pump or fan, an ultraviolet lamp, sensing and amplifying circuitry, a readout assembly containing a meter or display, controls, a power supply, and a battery.

The energy required to remove the outermost electron from the molecule is called the *ionization potential* (IP) and is specific for each chemical. The IP is measured in electron volts (eV). Several lamps are available with the PIDs. These are listed in Table 8.2.

As with the CGIs, these detectors are calibrated with a specific gas (usually isobutylene) and are only accurate when measuring the concentration of gases at or near the IP of the calibration gas. As the difference in

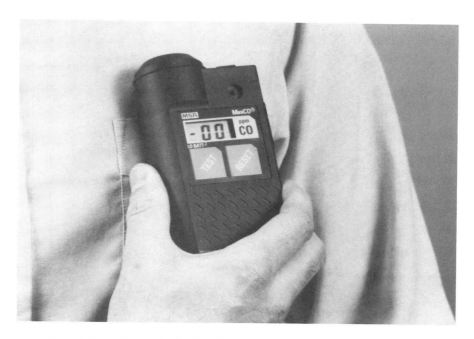

FIGURE 8.7 ■ A pocket-sized carbon monoxide indicator with digital readout. *Source:* Courtesy of MSA (Mine Safety Appliances Co.).

the IP of the detected gas and the calibration gas increases, the sensitivity to the detected gas is reduced.

These units are most useful in situations where the identity of the detected gas is known. In cases where the identity of the detected gas is unknown, the photoionization detector can give only a very gross indication of what the detected contaminant(s) may be or what the concentration of the detected gases are.

If only one chemical is present, the PID can be calibrated to that specific chemical. However, the PID will not quantitatively respond to a mixture unless all the chemicals in the mixture have the same IP.

The several manufacturers and users of these detectors have provided relative response factors for these instruments. Table 8.3 is a relative response table for the HNU with the 10.2-eV probe, calibrated to 10 ppm benzene.

The response to a gas or vapor may radically change when the gas or vapor is mixed with other materials. This could result in a substantial over-

242 ■ MONITORING

FIGURE 8.8 ■ Worker monitors drums on site for volatile organic compounds (VOCs) using a PID. *Source:* Courtesy of HNU Systems, Inc.

or understatement of the actual concentration of the contaminants. Figure 8.10 shows the HNU PI-101.

Flame Ionization Detectors (FID) The FID uses ionization as a detection method to detect flammable and/or toxic gases or vapors in much the same way as does the PID. The FID uses a hydrogen flame to ionize the vapors. The FID responds to virtually all compounds containing carbon–hydrogen or carbon–carbon bonds. When the vapors are burned, positive charged ions are created, which are then collected on a negatively charged electrode, producing a current which is converted to a concentration reading on the meter display. As noted above, FIDs respond only to

8.3 MEASURING INSTRUMENTS ■ 243

FIGURE 8.9 ■ Workers monitor drums at site for volatile organic compounds (VOCs) using a PID. *Source*: Courtesy of HNU Systems, Inc.

TABLE 8.2 ■ Ionization Potential of Several Commonly Used PIDs

Instrument	Available Lamps (IP)
HNU (NHU Systems, Inc.)	9.5, 10.2, 11.7 eV
Organic Vapor Meter (Thermo Environmental Instruments Inc.)	10.0, 11.8 eV
Microtip (Photovac, Inc.)	8.4, 9.5, 10.2, 10.6, 11.7 eV

TABLE 8.3 ■ Relative Responses of the HNU to Selected Organic Chemicals

Chemical	Photoionization Sensitivity (ppm)
p-Xylene	11.4
Benzene	10.0
Toluene	10.0
Styrene	9.7
Trichloroethylene	8.9
Isobutylene	7.0
Acetone	6.3
Methyl ethyl ketone	5.7
Vinyl chloride	5.0
Propylene	4.0
Hexane	2.2
Heptane	1.7
Methane	0.0
Acetylene	0.0

organic compounds. Thus, they do not detect inorganic compounds like chlorine, hydrogen cyanide, or ammonia.

The FID we will discuss below is the Foxboro organic vapor analyzer (OVA). This analyzer can operate in two modes, survey and gas chromatography. Several models of the Foxboro OVA are shown in Figure 8.11. In Figure 8.12 two workers are shown with the OVA 128 surveying the site for organic vapors.

In the survey mode the sample is ionized and the resulting current is translated on the meter to read concentration as total organic vapors. The OVA is factory-calibrated to methane. If the gas or vapor being detected is not methane, the meter reading will either understate or overstate the actual concentration of detected chemical, depending on whether the relative response of the detected chemical is higher or lower than the relative response of the calibrant gas. If the identity of the detected gas is known, then the actual concentration of detected chemical can be calculated. However, if the identity of the detected gas is not known, the actual concentration cannot be determined.

FIGURE 8.10 ■ HNU PI-101. *Source:* Courtesy of HNU Systems, Inc.

Table 8.4 gives information on the relative response of the OVA to some materials. Because the instrument is factory-calibrated to methane, all relative responses are given in percent, with methane at 100.

In the gas chromatographic mode it is possible to establish the identity of the unknown, detected, organic gases. This mode allows the OVA to separate the components of the sample and to match them to known chemicals by their retention time on the column (the time it takes for the components to elute from or leave the column). As the components leave the column, they flow into the detector, which can be connected to a strip chart which can record separate peaks for each component. This readout is called a *gas chromatogram*. The retention time is defined as the period of time that elapses between the injection of the sample and the time the sample components leave the column as a peak or peaks.

As with the PIDs, the FIDs respond differently to different compounds (see Table 8.4). Thus, the identity of the detected chemical(s) must be ascertained before any accurate determination of the concentrations is possible.

FIGURE 8.11 ■ Several models of the OVA family. In front is the TVA-1000. It is the first portable to incorporate both an FID and PID in a single package. Behind (from left to right) are the OVA 88, 108, and 128. *Source:* Courtesy of the Foxboro Company.

8.3.3 Classroom Exercise: Direct-Reading Instruments

Purpose This exercise will introduce the student to various direct reading instruments used to monitor for combustible gases, oxygen deficiency, and toxic gases. The student will learn to operate and sample using several different available instruments.

General Instructions

1. Prior to doing the actual exercise, list the monitoring capabilities of each instrument in the space provided for each on the correct exercise sheet. List such things as sensitivity, sensing range, sensing method (type), limitations, and so on.
2. Make sure that you use each instrument: that is, turn it on, adjust the zero (if necessary), operate, observe the meter readings, and record the results on the exercise sheet.
3. Prior to introducing the gases into the instruments, be sure that you know how to operate the instrument.

8.3 MEASURING INSTRUMENTS ■ **247**

FIGURE 8.12 ■ Several workers using the OVA 128 to survey the site for contamination. *Source*: Courtesy of the Foxboro Company.

4. Rotate from station to station (in groups) so that you will have an opportunity to operate each instrument. Each group will have approximately 20–25 minutes at each station, so work efficiently and quickly.
5. You will be using some or all of these instruments in the field exercises, so be sure to operate each one of them. These instruments are expensive ($2,000 to $7,500 each) and somewhat delicate, so please handle them carefully.
6. The calibration gases are expensive, so do not waste the gas. Filling the one-liter sampling bag should give you enough gas to do the test.
7. There will be four stations with an instrument at each station.

TABLE 8.4 ■ Relative Response of OVA Calibrated to Methane

Compound	Relative Response (%)
Methane	100
Ethane	90
Propane	64
n-Butane	61
n-Pentane	100
Ethylene	85
Acetylene	200
Benzene	150
Toluene	120
Methyl ethyl ketone	80
Carbon tetrachloride	10
Chloroform	70
Trichloroethylene	72
Vinyl chloride	35

Station A—photoionization detector

Station B—flame ionization detector

Station C—combustible gas/oxygen meter

Station D—toxic gas meter

8. The labeled sample bags and instruments will be at each station. The calibration gases will be at a central station, and all groups will share these gases.

Procedure

1. Go with your group to your first assigned station.
2. Take turns operating the instrument and get it prepared to monitor. Answer the general questions about the instrument listed on the exercise sheet.
3. One person will take sample bag A to the central gas station and fill the bag with gas A listed in your exercise.
4. Take gas in sample bag A back to your instrument and introduce this gas into the instrument.
5. Note what happens, record the results, and answer the questions on the exercise sheet.

6. A second person from your group will take sample bag B and go to the central gas station. Fill bag B with gas B listed in your exercise.
7. Return to your instrument, introduce the gas, record the results, and answer the questions on the exercise sheet.
8. A third person will repeat the above with sample bag C and gas C.
9. A fourth person will repeat the above with sample bag D and gas D.

TOTAL HYDROCARBON MONITORING EXERCISE

Exercise A, Station A: Photoionization Detector

General Questions:

1. Type of sensor _____

2. Sensitivity _____

3. Range _____

4. General chemicals it can sense _____

5. Advantages _____

6. Limitations _____

7. Special comments _____

Fill sample bag A with _____ ppm methane gas.

Do you expect this instrument to respond to this gas? _____

Why or why not? _____

Zero this instrument.

Introduce the methane in the instrument.

Record your reading. _____

Compare your reading with the stated calibration gas mixture found on the calibration gas cylinder.

Explain/comment _____

250 ■ MONITORING

_____

Why does the instrument respond/not respond to the gas? _____

_____

_____

Would you try 2.5% (50% LEL) methane gas on this instrument as you would a combustible gas monitor? _____

Why or why not? _____

_____

Fill up a second bag B with _____ ppm toluene or _____ ppm hexane, or whatever gas is furnished.

Do you expect this instrument to respond to this gas? _____

Why or why not? _____

Zero this instrument.

Introduce the gas in sample bag B to the instrument.

Record your reading. _____

Compare your reading to the actual concentration of the gas marked on the cylinder.

Explain/comment _____

Why does the instrument respond to this gas? _____

_____

Fill up a third bag C with _____ ppm carbon monoxide.

Do you expect this instrument to respond to this gas? _____

Why or why not? _____

8.3 MEASURING INSTRUMENTS ■ **251**

Zero the instrument.

Introduce the gas in sample bag C to the instrument.

Record your reading. _____

Compare your reading to the actual concentration of the gas marked on the cylinder.

Explain/comment _____

Why does the instrument respond to this gas? _____

_____

Fill up a fourth sample bag D with _____ ppm _____.

Do you expect this instrument to respond to this gas? _____

Why or why not? _____

Zero the instrument.

Introduce the gas in sample bag D to the instrument.

Record your reading. _____

Compare your reading to the actual concentration of the gas marked on the cylinder.

Explain/comment _____

Why does the instrument respond to this gas? _____

_____

Would you expect this instrument to respond to the following gas/vapors? if so, what levels?

Trichloroethylene _____

Hydrochloric acid _____

Comments _____

_____

TOTAL HYDROCARBON MONITORING EXERCISE

Exercise B, Station B: Flame Ionization Detector

General Questions:

1. Type of sensor _____

2. Sensitivity _____

3. Range _____

4. General chemicals it can sense _____

5. Advantages _____

6. Limitations _____

7. Special comments _____

Fill sample bag A with _____ ppm methane gas.

Do you expect this instrument to respond to this gas? _____

Why or why not? _____

Zero this instrument.

Introduce the methane in the instrument.

Record your reading. _____

Compare your reading with the stated calibration gas mixture found on the calibration gas cylinder.

Explain/comment _____

_____

Why does the instrument respond/not respond to the gas? _____

8.3 MEASURING INSTRUMENTS ■ 253

_____

_____

Would you try 2.5% (50% LEL) methane gas on this instrument as you would a combustible gas monitor?

Why or why not? _____

_____

Fill up a second bag B with _____ ppm toluene or _____ ppm hexane, or whatever gas is furnished.

Do you expect this instrument to respond to this gas? _____

Why or why not? _____

Zero this instrument.

Introduce the gas in sample bag B to the instrument.

Record your reading. _____

Compare your reading to the actual concentration of the gas marked on the cylinder.

Explain/comment _____

Why does the instrument respond to this gas? _____

_____

Fill up a third sample bag C with _____ ppm carbon monoxide.

Do you expect this instrument to respond to this gas? _____

Why or why not? _____

Zero the instrument.

Introduce the gas in sample bag C to the instrument.

254 ■ MONITORING

Record your reading. _____

Compare your reading to the actual concentration of the gas marked on the cylinder.

Explain/comment _____

Why does the instrument respond to this gas?

_____

Fill up a fourth sample bag D with _____ ppm _____.

Do you expect this instrument to respond to this gas? _____

Why or why not? _____

Zero the instrument.

Introduce the gas in sample bag D to the instrument.

Record your reading. _____

Compare your reading to the actual concentration of the gas marked on the cylinder.

Explain/comment _____

Why does the instrument respond to this gas?

_____

Would you expect this instrument to respond to the following gas/vapors? if so, what levels?

Trichloroethylene _____

Hydrochloric acid _____

Comments _____

TOTAL HYDROCARBON MONITORING EXERCISE

Exercise C, Station C: Combustible Gas/Oxygen Meter

General Questions:

1. Type of sensor _____

2. Sensitivity _____

3. Range _____

4. General chemicals it can sense: _____

5. Advantages _____

6. Limitations _____

7. Special comments _____

Fill sample bag A with _____ ppm methane gas.

Do you expect this instrument to respond to this gas? _____

Why or why not? _____

Zero this instrument.

Introduce the methane in the instrument.

Record your reading. _____

Compare your reading with the stated calibration gas mixture found on the calibration gas cylinder.

Explain/comment _____
_____

Why does the instrument respond/not respond to the gas? _____
_____
_____

MONITORING

Would you try 2.5% (50% LEL) methane gas on this instrument as you would a combustible gas monitor?

Why or why not?

Fill up a second sample bag B with _____ ppm toluene or _____ ppm hexane, or whatever gas is furnished.

Do you expect this instrument to respond to this gas? _____

Why or why not? _____

Zero this instrument.

Introduce the gas in sample bag B to the instrument.

Record your reading. _____

Compare your reading to the actual concentration of the gas marked on the cylinder.

Explain/comment _____

Why does the instrument respond to this gas?

_____

Fill up a third sample bag C with _____ ppm carbon monoxide.

Do you expect this instrument to respond to this gas? _____

Why or why not? _____

Zero the instrument.

Introduce the gas in sample bag C to the instrument.

Record your reading. _____

Compare your reading to the actual concentration of the gas marked on the cylinder.

Explain/comment _____

Why does the instrument respond to this gas?

_____

Fill up a fourth sample bag D with _____ ppm _____.

Do you expect this instrument to respond to this gas? _____

Why or why not? _____

Zero the instrument.

Introduce the gas in sample bag D to the instrument.

Record your reading. _____

Compare your reading to the actual concentration of the gas marked on the cylinder.

Explain/comment _____

Why does the instrument respond to this gas?

_____

Would you expect this instrument to respond to the following gas/vapors? if so, what levels?

Trichloroethylene _____

Hydrochloric acid _____

Comments _____

_____

TOTAL HYDROCARBON MONITORING EXERCISE

Exercise D, Station D: Toxic Gas Monitor

General Questions:

1. Type of sensor _____

2. Sensitivity _____

3. Range _____

4. General chemicals it can sense: _____

5. Advantages _____

6. Limitations _____

7. Special comments _____

Fill sample bag A with _____ ppm methane gas.

Do you expect this instrument to respond to this gas? _____

Why or why not? _____

Zero this instrument.

Introduce the methane in the instrument.

Record your reading. _____

Compare your reading with the stated calibration gas mixture found on the calibration gas cylinder.

Explain/comment _____

_____

Why does the instrument respond/not respond to the gas? _____

_____

_____

8.3 MEASURING INSTRUMENTS ■ 259

Would you try 2.5% (50% LEL) methane gas on this instrument as you would a combustible gas monitor? _____

Why or why not? _____

_____

Fill up a second sample bag B with _____ ppm toluene or _____ ppm hexane, or whatever gas is furnished.

Do you expect this instrument to respond to this gas? _____

Why or why not? _____

Zero this instrument.

Introduce the gas in sample bag B to the instrument.

Record your reading. _____

Compare your reading to the actual concentration of the gas marked on the cylinder.

Explain/comment _____

Why does the instrument respond to this gas? _____

_____

Fill up a third sample bag C with _____ ppm carbon monoxide.

Do you expect this instrument to respond to this gas? _____

Why or why not? _____

Zero the instrument.

Introduce the gas in sample bag C to the instrument.

Record your reading. _____

Compare your reading to the actual concentration of the gas marked on the cylinder.

Explain/comment _____

Why does the instrument respond to this gas?

Fill up a fourth sample bag D with _____ ppm _____.

Do you expect this instrument to respond to this gas? _____

Why or why not? _____

Zero the instrument.

Introduce the gas in sample bag D to the instrument.

Record your reading.

Compare your reading to the actual concentration of the gas marked on the cylinder.

Explain/comment _____

Why does the instrument respond to this gas?

_____

Would you expect this instrument to respond to the following gas/vapors? if so, what levels?

Trichloroethylene _____

Hydrochloric acid _____

Comments _____

8.3.4 Direct-Reading Colorimetric Indicator (Detector) Tubes

In evaluating hazardous waste sites and emergency response locations, the need often arises to determine the level of a specific chemical in air. Colorimetric indicator tubes can be used successfully to perform this task. They are usually calibrated to read in ppm or percent concentration for easy interpretation. There are tubes available for instantaneous and continuous sampling.

Theory, Application, and Use The interaction of two or more substances may result in chemical changes. These changes may result from the interaction of a colorless vapor or gas and a solid to produce a color change in the solid substance. Indicator tubes use this phenomenon to estimate the concentration of a gas or vapor in air.

Colorimetric tubes consist of a glass tube filled with an indicating chemical. The tube is inserted into the end of a piston cylinder or a bellows-type pump as shown in Figures 8.13, 8.14 and 8.15. A known volume of contaminated air is pulled at a predetermined rate through the tube. The contaminant reacts with the indicator chemical in the tube, producing a stain or color change whose length is proportional to the contaminant's concentration. Indicator tubes are usually chemical-specific. In other words, there are tubes for different gases—for example, chlorine tubes for chlorine gas, benzene tubes for benzene vapor, and so on. Some tube manufacturers do produce tubes that measure the concentration of groups of similar chemicals (e.g., aromatic hydrocarbons.)

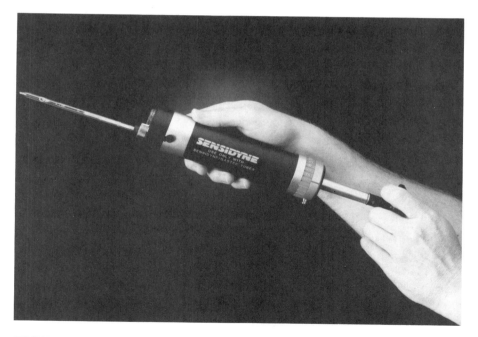

FIGURE 8.13 ■ Sensidyne/Gastec pump with analyzer tube.
Source: Courtesy of Sensidyne/Gastec.

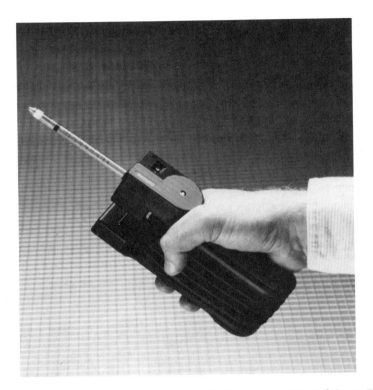

FIGURE 8.14 ■ Draeger accuro® bellows pump with a Dräger detector tube (accuro is a registered trademark of Drägerwerk AG). *Source*: Courtesy of National Draeger, Inc.

A preconditioning filter usually precedes the indicating chemical. Its purpose is to

- Remove moisture
- Remove contaminants (other than the one in question) that may interfere with the measurement
- React with the contaminant in question, to change it into a compound that reacts with the indicating chemical

Figures 8.16 and 8.17 show the Draeger bellows pump being used.

Advantages of Detector Tubes
- Simple to use
- Rapid and convenient
- Relatively quick response

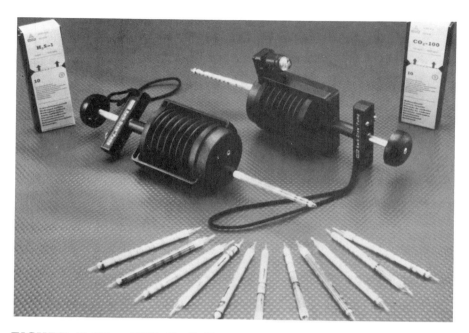

FIGURE 8.15 ■ MSA Kwik-Draw® pump and detector tubes.
Source: Courtesy of MSA (Mine Safety Appliances Co.).

- More than 100 types of tubes available
- Durable and rugged
- Portable
- Useful to determine the presence or absence of a suspect chemical, and to estimate the extent of a problem and whether more quantifiable measurements need to be taken

Disadvantages and Limitations

- In general, indicator tubes have an accuracy of plus or minus 25%. Some manufacturers report an error factor of up to 50% for some tubes.
- The chemical reactions involved in the use of the tubes are affected by temperature. Cold weather slows the reaction and thus the response time. To reduce this problem, it is recommended that in cold weather the tubes be kept warm until they are used (e.g., inside a coat pocket). Hot temperatures increase the reaction and can cause a problem by discoloring the indicator even if the contaminant is not

264 ■ MONITORING

FIGURE 8.16 ■ Worker using a Draeger detector tube system to measure soil gas at a landfill while wearing a Dräger powered air-purifying respirator. *Source:* Courtesy of National Draeger, Inc.

present. This can happen even in unopened tubes. Therefore, the tubes should be stored at a moderate temperature.
- The tubes indicate the concentration of contaminant only at the location the sample is taken and for the duration of the sampling.
- Many indicator tubes are prone to error due to the presence of interference gases/vapors. Read the instructions provided by the manufacturer regarding these interfering gases.
- Tubes are not available for all contaminants.

FIGURE 8.17 ■ **Closeup of a worker using a Draeger bellows-type pump.** *Source:* Courtesy of National Draeger, Inc.

- Remember indicator tubes have a self-life; therefore, stocking of tubes may be a problem.
- If the color change occurs rapidly (usually within a few seconds) and most or all of the tube contents change color, you can stop the test. This indicates that the concentration in the sample is above the highest reading on the tube. If available, obtain an indicator tube that can be used for the same chemical but with a wider range or insert another tube and take only half a pump stroke and check to see if the reaction is within the concentration range. Remember when recording the result you must multiply it by 2.

8.3.5 Qualitative Chemical Identification Exercise Using Detector Tubes

Introduction to the Use of the Pump and Tubes

1. Before sampling, check the pump performance by placing a sealed tube in the pump and testing according to the procedure described in the instruction booklet.
2. Take out the information sheet that comes in each box of tubes and read the instructions. Each different kind of tube has special instructions.
3. When ready to sample, break both tube ends by placing each of them in the special opening on the far end of the pump as indicated in the instructions. Be sure to hold the tube firmly between your fingers and right next to where you are making the break. This allows for a straighter and cleaner break of the tube and helps prevent the tube from breaking further up the tube and cutting your hand.
4. Push the handle of the pump in so that the red circle is just visible.
5. Place the tube in the rubber collar so that the arrow on the tube points toward the pump and the lowest numbers on the tube are furthest from the pump. Remember that the air will be moving through the tube into the pump once you pull the handle.
6. Move to the place you want to sample, and place the tube into the area of concern. Line up red dots on the piston shaft and the pump body and pull out the piston with one steady pull. Hold the pump in place for the required time as indicated on the information sheet. After the time is up, twist the pump handle. It should not move in but have to be pushed back in place. If it jumps in, you did not wait long enough and the sample is not valid. You must then repeat the testing.
7. The Gastec pump (model 7010790-1) that you will be using has a flow indicator. It is located at the end of the pump next to the rubber detector tube holder. The indicator has a flat red circle inside the clear cylindrical plastic casing of the indicator. The red indicator moves to a position flush with the pump body so it cannot be seen from the side view when the pump handle is pulled out. It returns slowly to its original position as the required volume of sample has moved through the tube.
8. If little or no color change is observed, take another stroke or pull of the pump, making sure that the tube is at the same place or area of

sampling. Consult the specific tube information sheet on how many additional pump strokes should be taken and how to read the tube. Most tubes are direct reading with one stroke. If you use two strokes, divide the reading in half to get the correct reading.

9. If the stain rapidly discolors the entire tube, you probably cannot get a good reading on the tube. Use a tube with wider concentration range if available. If the stain slowly discolors the entire tube, try a second tube and use a half stroke rather than a full stroke.
10. Remove the tube from the pump and take your readings. Read the entire length of the stain even if part of the stain is lighter in color.
11. Record the reading and other information as follows:

Sample	Date	Time	Tube #	Area or Location of Sampling	Special Conditions	Pump Strokes	Final Reading

Instructions for the Detector Tube Exercise

1. *Equipment needed*
   a. Determine which chemicals you will encounter. For this exercise you will be given three known chemicals to test (Known A, B, and C).
   b. Determine, select, and procure the tubes you will require for the exercise.
   c. Procure a Gastec pump kit.
2. *Procedure*
   a. Obtain an information sheet for each different kind of detector tube to be used. These sheets can be found in each new box of tubes and in the Gastec detector tube *Blue Book of Tube Instructions*.
   b. From these sheets note the following:
      i. The color change from the original color when it is exposed to the primary intended chemical—that is, acetone tube when exposed to acetone.

ii. Note which are the interference chemicals and the color change expected when the tube is exposed to the interference chemicals; that is, toluene tube turns pale brown when exposed to hexane.
c. From these individual detector tube information sheets, fill in the Color Change/Chemical Interference Chart as follows:
   i. Place the name of the known chemical A in the blank space labeled A at the top of the Chemical Interference Chart. Do the same for B and C.
   ii. Place the name of the detector tube and its number you will be using in the blank space in the far left column with the tube for Known A being in the space marked a, the tube for Known B in the space marked b, and so on.
   iii. Place the color change that tube a turns when exposed to its primary chemical—Known A in the blank number 1.
   iv. Place the color change that tube a turns when exposed to interferant chemical B (Known B) in blank 2.
   v. Place the color that tube a turns when exposed to interferant chemical C in blank 3, and so on.
   vi. Place the color change that tube b turns when exposed to its primary chemical B—Known B in the space marked 5.
   vii. Place the color change that tube b turns when exposed to interferant chemical A in the space marked 4.
   viii. Place the color change that tube b turns when exposed to interferant chemical C in space 6.
   ix. Complete the balance of the chart in a like manner for all the other chemicals used in this exercise.
d. Look over the chart, noting the various color changes. Try to determine if there is a logical scheme you can use to separate out or identify one or more of these chemicals from the others using the detector tubes with their various color reactions. The identified standards are provided to help with the identification process if you want to use them.
e. The purpose of this exercise is to gain experience in using the detector tube pump and tubes and to identify at least one of the chemicals using this scheme.
f. Using the pump and the various detector tubes, test each unknown and determine which of the known chemicals A, B, C, and so on, is in each of the unknown jars—Unknown F, G, H, and so on.
g. Record your results in the Color Change/Chemical Interference Chart.

General Exercise Information and Problem Solving Suggestions

1. *General information to help in your strategy*
   a. All chemical standards (knowns) are correctly identified and are the pure chemicals as labeled.
   b. All unknowns are pure samples; there are no mixtures of these various chemicals in the unknowns.
   c. All unknowns are taken from the group of chemicals used as the standards or Knowns (A–C) in this exercise.
   d. Each of the knowns may not be represented as an unknown. The unknowns could be all the same chemical or all different ones or any combination of them. Therefore, each unknown should be tested on its own.
   e. Have each person in the group take a turn at using the pump and detector tubes.

2. *Problem-solving suggestions*
   a. Develop a system of identifying and keeping track of the tubes.
   b. Plan what you are going to do before you start doing it. Develop a strategy.
   c. Use as few tubes as possible and work as efficiently as possible. Remember you may be under time constraints and/or equipment limitations; for example, you may be wearing a SCBA with a limited air supply.
   d. Some of the sample concentrations are over their TLV, so we want as little exposure to you as possible. Open the jar just far enough to insert the detector tube. Pull the pump back promptly. Note the color change. As soon as you observe a significant color change (usually within a few seconds), withdraw the tube from the jar and close the lid tightly. You do not have to wait the prescribed time. This is a qualitative exercise, not a quantitative one, so the full 100-cc sample is not required. However, if you do not observe a color change right away, leave the tube exposed in the jar for the full time. You may even want to take a second sample pull to make sure there is no reaction. If you observe that no reaction has occurred, you can assume that the tube is not responding to the contents of the jar and that the specific chemical is not present at detectable levels.
   e. Do not reuse a tube (except when taking another pull on the pump when testing the same sample). Reuse of tubes, even if no reaction is observed, is a very poor practice and should not be allowed. Reactions

can occur in the tube which are colorless and can affect subsequent tests. The price of the tube does not warrant the cost of your time and problems associated with inaccurate results.

f. Be prepared to present your group result after you complete the exercise. *Do not* share your results with the other groups. Each group is judged on its own merits.

QUALITATIVE DETECTOR TUBE EXERCISE

Equipment needed
1. Detector tube pump kit
2. Detector tubes for the Knowns (standards) A, B, and C.
   a. _____
   b. _____
   c. _____
3. Standards
   A. _____
   B. _____
   C. _____

Color Change/Chemical Interference Chart

Detector tubes	A.	B.	C.
a.	1	2	3
b.	4	5	6
c.	7	8	9

Unknowns
   F. _____
   G. _____
   H. _____

8.4 PERSONAL AIR SAMPLING

Another means of monitoring a site is by sampling the immediate surroundings of persons working in the contaminated environment. Working on hazardous waste sites requires that individuals as well as the general environment be continuously monitored for dangerous gases, vapors, and particulates. One of the methods to accomplish this is to sample the immediate surroundings of persons working in the hazardous environment. Placing sampling devices on affected workers in their breathing zone areas can aid in determining and verifying contamination levels previously obtained by direct-reading equipment.

8.4.1 Choosing Sampling Methods

The atmosphere may be sampled during a hazardous waste cleanup operation to identify and quantify any gases, vapors, or particulates to which workers can be exposed. Such information can be obtained by two methods:

- Area sampling, which involves the placement of collection devices within designated areas and operating them over specific periods of time.
- Personal sampling, which involves the collection of samples from within the breathing zone of an individual, sometimes by the individual wearing the sampling device. Figure 8.18 shows several personal pumps ready for use.

Once the sampling method has been selected, the type of sampling desired must be determined. Prevailing conditions, the scope of site operations, and the intended use of the information dictate the type of sample collected.

Instantaneous or grab-type samples are collected for short time periods. They are useful in examining stable contaminant concentrations or peak levels for short durations. Instantaneous samples may require highly sensitive analytical methods due to the small sample volume collected.

Integrated samples are more typical of on-site measurements. They are collected when the sensitivity of the analytical method requires minimum sampling periods or volumes, or when comparison must be made to an 8-hour, time-weighted average threshold limit value (TLV) or the OSHA standard, permissible exposure limit (PEL).

FIGURE 8.18 ■ Several sampling pumps ready for use in the field to determine worker exposure. The pump clips on to the belt and the plastic tube runs behind the back and clips onto the collar where the sample enters and is trapped. The pumps operate for a complete shift or day. The pump output is used to determine the worker's time-weighted average exposure (TWA).

Two types of sampling systems are used for the collection of samples:

- Many of the new active samplers mechanically move contaminated air through a sensor. These instruments give the operator a continuous readout of the contaminant the instrument can sense and/or a long-term average or TWA. Other active samplers mechanically move contaminated air through a collection medium which traps the contaminants for subsequent analysis.
- Passive samplers rely on natural rather than mechanical forces to collect samples. Passive samplers are classified as either diffusion or permeation devices, according to their principle of operation.

The sampling system chosen depends on a number of factors, including

- Instrument or system efficiency
- Operational reliability
- Ease of use and portability
- Availability of instrument and component parts
- Information or analysis desired
- Personal preference

Active Samplers Most active sampling systems mechanically collect samples on or into a selected medium. The medium is then analyzed in the laboratory to identify and quantify the contaminant(s) collected. Such a system typically consists of the following components:

- An electrical pump containing a flow regulator to control the rate of movement of the contaminated air and a flow monitor to indicate that rate.
- The appropriate sampling medium and a sampling container designed for that medium. These items depend largely upon the contaminants to be sampled and the selected sampling method.
- Flexible, nonporous, inert tubing linking the sampler to the pump. Integrated samples are usually collected for a given time period and at known, fixed flow rates. Thus, pump calibration and accurate time measurement are critical to the proper interpretation of the data collected by active systems.

Sampling Pumps Active sampling systems, including those which contain sensors or collection media, typically rely on electrical pumps to mechanically induce air movement. The most practical electrical sampling pumps are powered by rechargeable batteries and can operate continuously at constant flow rates for at least 8 hours. Typically, they are compact, portable, and quiet enough to be worn by individuals when monitoring personal exposures (see Figure 8.19). The type of portable pump selected for a collection medium system is generally determined by such factors as the physical properties of the contaminant, the collection medium, and the collection flow rates specified by the analytical method used.

On the other hand, the pump–sensor combination system is selected by the sensor required to monitor the specific contaminant present or presumed to be present at the site. These new gas/vapor detectors contain one to four different sensors that can measure accurately at the ppm level a variety of contaminants.

274 ■ MONITORING

FIGURE 8.19 ■ Demonstration of a sampling pump with the collection tube running behind the demonstrator's back and clipped to his collar.

Sample Collection Devices

Gases/Vapors. Active samplers for gaseous and vapor contaminants make use of a variety of collection media, including solids, liquids, long-duration colorimetric tubes, and sampling bags. All require a pump to move the contaminant containing air to and through the collection medium.

Solid Sorbents. These materials are the class of media most widely used in hazardous materials sampling operations. They are collected by adsorption and are often the media of choice for insoluble or nonreactive gases or vapors.

- The two most widely used solid sorbents are activated charcoal and silica gel:

    Activated charcoal has the broadest range of collection efficiencies. The highest efficiencies are for organic gases with boiling points above $0°C$, and the lowest efficiencies for organic gases with boiling points below $-150°C$.

Silica gel is the next most widely used solid sorbent. It will adsorb in decreasing order of efficiency the following: water, alcohol, aldehydes, ketones, esters, aromatic compounds, olefins, and paraffins.

A number of other synthetic sorbents are available for specific gas or vapor contaminants or groups of contaminants. These are the following:

Solid Sorbent	*Adsorbed Chemical or Group*
Floricil	Polychlorinated biphenyls
Chromosorb 101	*bis*-Chloromethyl ether
Chromosorb 104	Butyl mercaptan
Porpak Q	Furfuryl alcohol and methyl cyclohexanone
XAD-2	DDVP, Demeton, ethyl silicate, nitroethane, quinone, tetramethyllead (as lead)

The solid sorbent materials are commonly purchased as sampling tubes (see diagram below) which are used in conjunction with an electrical sampling pump.

Sampling tube containing two sections of sorbent media

Liquid Adsorbents. These are used with impingers or bubblers (see diagram below) and powered pumps to collect soluable or reactive gases. Liquid adsorbents commonly used include the following:

Liquid Adsorbent	*Adsorbed Chemical or Group*
$0.1N$ H_2SO_4	Bases and amines
$0.1N$ NaOH	Acids and phenols
$0.1N$ HCl	Nickel carbonyl
Alkakline $CdSO_4$	Hydrogen sulfide
Methyl blue	Hydrogen sulfide
Nitro reagent (4-nitropyridylpropylamine in toluene)	Diisocyanates

0.3N H$_2$O$_2$ Sulfur dioxide
0.1% Aniline   Phosgene
1% NaHSO$_2$ Formaldehyde
Distilled water   Acids and bases

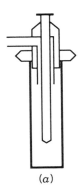 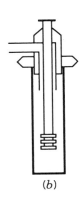

Impinger (a) and bubbler (b)

Long Duration, Direct-Reading Colorimetric Tubes. These are generally used in the 2- to 8-hour range, and they may be used for time-weighted average sampling. Unlike their short-term counterparts, long-duration tubes require an electric pump rather than a hand-operated pump.

Sample Bags. These offer alternatives in the collection of instantaneous grab samples of gases and vapors, and they are particularly useful when representative samples are desired. Sampling is performed by connecting the bag inlet valve with flexible tubing to the exhaust outlet of a sampling pump. The bag inlet valve is opened, the pump is turned on, and the sample is collected. Once the sample is collected, the bag contents may be sampled or directly emptied into an analytical instrument, such as a gas chromatograph, or tested by colorimetric tube.

Particulates. Airborne particulates or aerosols include both dispersed liquids (mists and fogs) and solids (dusts, fumes, and smoke). The most common method of sampling particulates is to trap them on filters using active systems to collect the samples. Particulate collectors are generally used to selectively gather samples on the basis of particle size. Particulate samplers typically have an air inlet and a membrane or fiber filter on which the particulates are collected. A preselector may also be included prior to the filter if the particulates are to be classified by size.

Cyclones or centrifugal separators are commonly used to separate and collect particles small enough to enter the respiratory system. Cyclones are typically used as personal respiratory dust monitors.

Connected to most preselectors is a cassette containing a desired filter material. Typically, these cassettes are molded out of transparent polystyrene plastic. They consist of two or three stacked sections, the number depending on the contaminant and the collection method.

Two types of filter materials are used:

- *Fiber filters* are composed of irregular meshes of fibers forming openings or pores whose diameter is 20 micrometers or less. A number of fiber filters are available. The two with perhaps the greatest application to hazardous materials operations are *cellulose* and *glass*. Of the two, cellulose is the least expensive and is available in a variety of sizes. Its greatest disadvantage is its tendency to absorb water, thus creating problems in weighing. For this reason, glass filters are finding more applications.
- Membrane filters are microporous plastic films. This group includes such materials as cellulose triacetate, polyvinyl chloride, teflon, polypropylene, nylon, and silver (Table 8.5). These filters have an ex-

TABLE 8.5 ■ Filter Media for the Collection of Airborne Particulates

Filter Medium	Representative Application
Cellulose ester, 0.45-micrometer pore	Metal fumes, acid mists
Cellulose ester, 0.80-micrometer pore	Asbestose, metal fumes, fibers, chlorodiphenyls (54% chlorine)
Fibrous glass	Total particulates, oil mists, pesticides, and coal, tar, and pitch volatiles
Polyvinyl chloride, 5.0-micrometer pore	Weight analysis, hexavalent chromium
Silver membrane	Total particulates; coal, tar, and pitch volatiles; free crystalline silica
Teflon	High-temperature applications

tremely low mass and ash content. Some are completely soluble in organic solvents, enabling the concentration of collected particulates into small volumes for later analysis.

8.4.2 Sampling Equipment and Procedures

Passive Samplers Quantitative passive samplers have been available since the early 1970s. The major advantage of passive dosimeters is their simplicity. These small, lightweight devices do not require a mechanical pump to move the contaminants through the collection medium. Thus, calibration and maintenance are reduced or eliminated, although the sampling periods must still be accurately measured. The same sources of error exist for both passive and active samplers. These are interpretation of results and the effects of temperature and humidity.

The passive samplers commercially available apply to gas and vapor contaminants only. These devices are used mainly as personal exposure monitors, although they have some usefulness in area monitoring. Passive samplers are divided into two groups, primarily on the basis of how they are designed and operated. Figures 8.20 and 8.21 show passive diffusion samplers.

- Diffusion samplers function by the passive movement of contaminant molecules through a concentration gradient created within a stagnant layer of air sandwiched between a contaminated atmosphere and the indicator material. Some diffusion samplers can be read directly, as are colorimetric detector tubes, while others require laboratory analysis similar to that performed on solid sorbents.
- Permeation samplers rely on the natural permeation of a contaminant through a membrane. Permeation samplers are useful in picking out a single contaminant from a mixture of airborne contaminants due to the selective permeability of chemicals through specific membranes. As with diffusion samplers, some passive permeation samplers may be direct-reading, while others may require laboratory analysis.

Calibration Atmospheric sampling systems must be accurately calibrated to a specific flow rate if the resultant data are to be correctly interpreted. Flow rate calibration of the electric pump in active systems is

FIGURE 8.20 ■ Closeup of a 3M™ 3500 Series mercury vapor monitor clipped to the collar of a shirt. *Source*: Photo compliments of 3M Occupational Health and Environmental Safety Products.

important in achieving the constant flow rates often specified in standard analytical methods. As a minimum, an active sampling system should be calibrated prior to use and following a prescribed sampling period. The overall frequency of calibration depends upon the general handling and use a sampling system receives. Pump mechanisms should be recalibrated after they have been repaired, when newly purchased, and following any suspected abuse.

However, passive sampling systems, because of their simplicity in design and principles of operation, require no formal calibration.

Personal Sampling Plan   It is not easy to write a plan for sampling personnel engaged in hazardous waste site operations, for the following reasons:

FIGURE 8.21 ■ **Closeup of a 3M™ 3600 Series organic vapor monitor clipped to the collar of a shirt.** *Source:* Photo compliments of 3M Occupational Health and Environmental Safety Products.

- It is often difficult to decide what to sample for.
- Many workers move around on the site, and some have many different tasks.
- Keeping track of sampling data in an efficient and informative manner requires attention and organization.

Currently there are only a few substances for which a full standard exists. These include lead, asbestos, chromium, acrylonitrile, and several others. No required sampling method exists for other substances, and if air concentrations do not exceed the permissible exposure limits, personnel monitoring is not required. If the PELs are exceeded on site, the following guidelines may be used:

- Provide the required chemical protective clothing and respiratory protection.
- Sample workers who are known to be exposed and/or who are working in heavily contaminated locations.

- If the hazardous locations on-site have not been determined, sample all workers or work locations routinely.
- Group workers by exposure levels.
- Sample some members of each group every week.
- Change the sampling plan as conditions change on-site.

Monitoring Strategies In general, personal air monitoring should be conducted whenever

- Work begins on a different portion of the site
- New contaminants are encountered
- A different type of operation is initiated
- Drums or other containers are being handled
- Obvious areas of liquid contamination are encountered

Classically, personal air monitoring is done using personal samples—that is, samples collected over the course of the work day taken near the breathing zone (a two-foot sphere encompassing the head) of the worker.

Given the sporadic nature of hazardous waste site exposures and the need to know immediately whether air contaminants are present, direct reading instruments are used instead. Generally, these are short-term and are collected during periods such as soil sampling, drilling, and drum sampling when contaminants are most likely to be present.

A better approach to monitoring is to integrate and use both methodologies when appropriate. For more complete information about air contaminants, spot measurement obtained with direct-reading instruments must be supplemented by continuous collecting and analyzing of long-term (full-day) air samples. Both personal and area sampling are necessary. For example, to assess the potential for "contaminant drift," air sampling devices equipped with the appropriate collection media are stationed at various locations throughout the area. These samples provide air quality information for the period of time they operate, and they can be set up to be specific for types of contaminants and concentration required for the specific site operations. It is important to review the data regularly and make adjustments in the types of analyses when required. In addition to personal air samplers, area sampling stations may also include direct-reading instruments equipped with recorders operating continuously. Area sampling stations should be stationed in various locations, including:

Upwind. Because many sites exist near industries or highways that generate air pollutants, area samples must initially be taken upwind of the site to establish background levels of air contaminants.

Support Zone. Periodic area samples must be taken near the command post or other support facilities to ensure that they are in fact located in the clean area and that the area remains clean throughout operations at the site.

Contamination Reduction Zone (decontamination zone). Air samples should be collected along the decontamination line to ensure that decontamination workers are properly protected and that on-site workers are not removing their protective gear in a contaminated area.

Exclusion Zone. The exclusion zone presents the greatest risk of exposure to chemicals and requires the most air sampling. The location of sampling stations should be based upon hot-spots detected by direct-reading instruments, the types of substances present, and the potential for contaminants to become airborne. The data from these stations, in conjunction with intermittent walk-around surveys with direct-reading instruments, can be used to verify the selection of proper levels of worker protection and exclusion zone boundaries, as well as to provide a continual record of air contaminant concentrations.

Downwind. One or more area sampling stations are located downwind from the site to indicate if any contaminants are leaving the site. If there are indications of airborne hazards in populated areas, additional samplers should be placed downwind.

Organic Gases and Vapors. The sequence for monitoring organic gases and vapors consists of several steps:

- Determine total background concentrations.
- Survey site for areas of high contamination (hot-spots).
- Collect on-site samples.
- Identify specific contaminants.

Background Concentrations. Determine background readings of total organic gases and vapors, using a PID or FID. These are usually made upwind of the site in areas not expected to contain air contaminants. If industries, highways, or other potential sources contribute to concentrations on-site, these contributions called background should be determined.

Concentrations On-Site. The on-site area is monitored using a PID or FID to determine total gas and vapor concentrations, measured at both ground and breathing zone levels. The initial walk-throughs are to determine general ambient concentrations and to locate higher-than-ambient concentrations (hot-spots). Transient contributors on-site (for example, exhaust from engines) should be avoided. Concentrations are recorded and plotted on a site map. Additional direct-reading instrument monitoring is then done to thoroughly define any hot-spots located during the survey.

Area Samples. Sampling stations are located throughout the site. The number and locations depend on many factors, including hot-spots, active work areas, activities with a potential for high concentrations, and wind direction. As a minimum, stations should be located on a boundary line downwind on a clean off-site area (control or background station) and in the exclusion zone. As data are accumulated, the location, number of stations, and frequency of sampling can be adjusted. Area surveys using direct-reading instruments are continued routinely two to four times daily. These surveys are to determine general ambient levels, as well as levels at sampling stations, hot-spots, and other areas of site activities. As information is accumulated on airborne organics, the frequency of surveys can be adjusted.

Identification of Specific Contaminants. If the identities of airborne contaminants are narrowed down to a few organic gases and vapors, it is better to set up a program to specifically monitor for them. This may be accomplished by using direct-reading instruments that sense these contaminants or the specific detector tubes, charcoal tubes, or other collection media depending on the type of contaminant and method.

Particulate and Inorganic Gases and Vapors. Sampling for particulates is not done routinely. If these types of air contaminants are known or suspected to exist, a sampling program is instituted for them. Incidents where these contaminants might be present are fires involving pesticides or chemicals, incidents or activities involving heavy metals, arsenic, or cyanide compounds, or remediation and/or mitigation operations that create dust (from contaminated soil and the excavation of contaminated soil).

Carcinogens. If known carcinogens such as polychlorinated biphenyls (PCBs), polynuclear aromatics (PNAs), coal tar, and various petroleum products such as creosote oil, pitch, anthracene oil, and inorganic metals such as chromium, nickel, beryllium, cobalt, and so on, are present at the site, personal sampling should always be done to document exposures, regardless of how long concentrations are expected to be present.

Meteorological Considerations Meteorological information is an integral part of an air surveillance program. Data concerning wind speed and direction, temperature, barometric pressure, and humidity, singularly or in combination, are needed for

- Selecting air sampling locations
- Calculating air contaminant dispersion
- Calibrating instruments
- Determinating populations or communities at risk of environmental exposure from airborne contaminants

Knowledge of wind speed and direction is also necessary to effectively place air samplers. In source-orientated ambient air sampling particularly, samplers need to be located downwind (at different distances) of the source and others placed to collect background samples. Shifts in wind direction must be known and samplers relocated or corrections made for the shifts. In addition, those in charge of air monitoring who are utilizing computerized atmospheric simulation models for predicting contaminant dispersion and concentration levels need wind speed and direction to generate an accurate prediction.

At minimum, a site should be equipped with a wind sock, accurate thermometer, and barometer. The choice of how information is obtained depends upon the availability of reliable data at the location desired, resources needed to obtain meteorological equipment, accuracy of the information needed, and the use of the information.

8.5 SUMMARY

What Can Be Monitored in the Air? Generally, air monitoring is done to determine the presence or potential of four atmospheric hazards:

- Toxic chemicals
- Fire and explosion hazards
- Oxygen deficiency
- Oxygen enrichment

Toxic Chemicals Detecting the presence and measuring the concentration of toxic chemicals is of utmost importance to the site worker. However, the ability to determine the specific hazard by monitoring the air will be limited to the capabilities of the monitoring instrument used.

Fire and Explosion Hazards Determining if there is a possibility for fire or explosion is critical during many emergencies and site activities. Explosive atmospheres develop from evaporation of flammables, gas leaks, and accumulation of dust. Potentially flammable atmospheres must be monitored frequently in accordance with the site safety plan. Protective clothing and respirators that protect the first responder from toxic hazards provide little, if any, protection against fire or explosions.

Oxygen-Deficient Atmospheres Without an adequate concentration of oxygen in the air, the site worker's health is immediately at risk (IDLH, Immediately Dangerous to Life and Health). The air normally contains approximately 21% oxygen; below 19.5% a person may become lethargic and may eventually lose consciousness. Confined spaces, such as tanks and pits, may contain an oxygen-deficient atmosphere. Potentially oxygen-deficient atmospheres (below 19.5%) must be monitored frequently.

Because respirators, which only purify the air (APR), do not protect the wearer from oxygen-deficient atmospheres, air must be supplied. If the oxygen concentration is below 19.5%, it is required that the worker be supplied an airline respirator or a SCBA.

Oxygen-Enriched Atmospheres Too much oxygen can cause health problems and increase the risk of fire or explosion. Some chemicals, known as *oxidizers*, release oxygen and cause an oxygen-enriched environment. Other situations where oxygen may be present at unusually high levels include where oxygen is being used in the process. An oxygen-enriched atmosphere is defined as one containing more than 23.5% oxygen and can cause materials that wouldn't ordinarily ignite to ignite and burn vigorously. Potentially oxygen-enriched atmospheres must be monitored frequently.

An oxygen-enriched atmosphere poses a threat of explosion, especially if flammable materials are present. Special procedures are necessary for testing the area, entering the area, and working in the area. These activities should be done only if they are permitted by the site safety plan.

8.5.1 Types of Air Monitoring

Area Monitoring Air samples may be collected by placing monitoring devices directly on the worker (personal monitoring) or by placing a monitoring device in the area where people may work (area monitoring). Monitors (personal or area) with alarms can be set up in a work area to alert workers to unusual or unexpected concentrations of substances in the air.

Real-time monitoring is a type of area monitoring which provides a direct reading of air contamination at the moment it is being used. Other types of area monitors collect samples which are sent away for analysis by a laboratory.

Personal Monitoring Personal monitoring is done to determine the quality of the air the worker is breathing or would breathe if not protected. Personal air samples are usually collected by placing a battery-operated air pump on the wearer's belt and clipping a collection tube or filter on his/her collar near the nose, an area known as the *breathing zone*. Air from the environment is pulled into the collection tube or filter where the contaminants are trapped. The collection tube or filter is sent to a laboratory for analysis.

Another method of collecting personal air samples is through the use of a passive dosimeter. This device is a chemically sensitive badge clipped to the worker's collar which collects a sample without using a pump.

Advantages of Personal Air Monitoring
- It is the most accurate measurement of worker's actual exposure.
- The results can be converted to a TWA and compared with the OSHA-PEL and ACGIH-TLV levels.
- It can be used during site activities to document exposure throughout the project.

Disadvantages of Personal Air Monitoring
- Some methodologies require laboratory analysis of the sample, which may take 1–14 days.
- It provides no data concerning peak or ceiling exposures if collected over several hours.
- It generally requires knowledge of the specific chemical in the air.
- It requires preparation so the equipment is ready when needed for monitoring site personnel.

Real-Time Monitoring Real-time monitoring provides an immediate measurement of substances in the air. It can be done with a variety of equipment. The equipment selected at any site will normally depend upon the potential hazards present. The real-time monitors are often referred to as direct-reading instruments. Various types can detect gases, vapors, dusts, flammable atmospheres, oxygen, radiation, heat and noise.

Advantages of Real-Time Monitoring
- It allows information to be immediately available at the site.
- It is available for a wide range of potential hazards.
- It measures chemicals that might cause acute health effects and ILDH situations.
- It allows workers to identify potentially high levels of toxic and flammable materials.
- It helps determine whether the atmosphere within a confined space is safe for entry (not oxygen-deficient, flammable, or toxic).

Disadvantages of Real-Time Monitoring
- It may not be sensitive to hazards present or sensitive enough to detect low levels of contaminants.
- Many of these instruments cannot identify a specific contaminant or distinguish one contaminant from another.
- It may give an inaccurate reading because of background levels or the presence of chemicals other than the one being sampled.
- It may require frequent calibration and/or factory maintenance.

8.5.2 Selecting Monitoring Equipment

Project management is responsible for selecting equipment appropriate for anticipated activities and potential hazards. Manufacturers should provide information about the uses and limitations of their equipment.

- The unit should be intrinsically safe (that is, it will not produce sparks that could trigger an explosion). Check the label and the manufacturer's instrument manual to make sure.
- Most instrument are designed to monitor only one contaminant. There are *no* instruments which can monitor *all* toxic substances.
- Equipment should be easy to operate in the field under changing conditions.
- Instruments should operate properly at temperatures which are anticipated during response activities.

- Training should be available which gives workers responsible for air monitoring a chance to routinely practice with the equipment.

REVIEW QUESTIONS

1. What are two types of monitoring that might be done during site activities?
2. Is a STEL or TWA generally applicable to site worker exposures? Why or why not?
3. Match the following:
   Gas/vapor   f/cc
   Fiber   mg/m^3
   Dust/mist   ppm
4. What instruments are used to test air in a confined space?
5. The combustible gas meter gives a reading of 0%. What are the possible reasons for this reading?

REFERENCES

American Conference of Governmental Industrial Hygienists, *Air Sampling Instruments for Evaluation of Atmospheric Contaminants*, 7th ed., ACGIH, Cincinnati, OH, 1989.

EPA, *Standard Operating Safety Guides*, U.S. Government Printing Office, Washington, D.C., 1984.

Foxboro, *Product Specifications, Model OVA 128, Century OVA Portable Organic Vapor Analyzer*, Foxboro Company, Foxboro, MA, 1985.

Foxboro, *Product Specifications, Model TVA-1000, Century OVA Portable Toxic Vapor Analyzer*, Foxboro Company, Foxboro, MA, 1994.

HNU, *Instruction Manual, PI 101, Portable Photoionization Analyzer*, HNU Systems, Inc., Newton, MA, 1986.

Levine, S., and W. Martin, eds., *Protecting Personnel at Hazardous Waste Sites*, Butterworth, Stoneham, MA, 1985.

REFERENCES

3M Occupational Health and Environmental Safety Division, *3M™ Monitor Selection Guide*, 3M, St Paul, MN, 1995.

NIOSH/OSHA/USCG/EPA, *Safety and Health Guidance Manual for Hazardous Waste Site Activities*, U.S. Government Printing Office, Washington, D.C., 1985.

CHAPTER 9

DECONTAMINATION

The process of removing or neutralizing contaminants that have accumulated on personnel and equipment is termed *decontamination*. This process is critical to health and safety at hazardous waste because it protects workers from hazardous substances that may contaminate and eventually permeate the protective clothing, respirators, and other equipment used on site. Decontamination or "decon" also protects personnel by minimizing the transfer of hazardous materials into clean areas.

CHAPTER OBJECTIVES

When you have completed this chapter, you will

- Know the goals, importance, and components of a good decon plan
- Differentiate between physical and chemical decontamination methods and the appropriate use of each
- Be able to describe procedures for decontaminating personnel, clothing, tools, and vehicles
- Be aware of the effectiveness of decon evaluation techniques

9.1 WORK ZONES

It is critical to prevent movement of contaminated materials and contaminants from the exclusion or contaminated zone. An important procedure for accomplishing this is to establish *work zones* as shown in Figure 9.1.

Three work zones should be established:

- The *hot zone* is the contaminated area, where the remediation activities are occurring.
- The *warm zone or contamination reduction zone (CRZ)*, which surrounds the hot zone, is the area where decontamination occurs.
- The *cold zone or support zone*, the area free of contamination, is where support activities occur.

Primary Activities in Each Work Zone Different activities take place in each zone. These activities and the personnel required to conduct the activities should be restricted to appropriate zones.

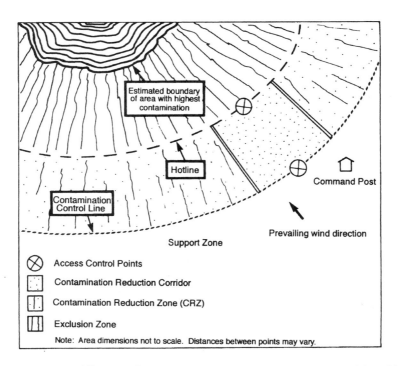

FIGURE 9.1 ■ **Site work zones.** *Source:* NIOSH, *Safety and Health Guidance Manual for Hazardous Waste Site Activities,* 1985.

Hot Zone. The remediation activities are ongoing in this area. The size of the zone is determined by the extent of the contamination on site, characteristics of the site, and access points. The "hotline" is the outer boundary, which should be clearly marked with fences, hazard tape, lines, signs, or ropes. Further subdivision of the area may be necessary, depending on the hazard and incompatibility of materials. The level of PPE necessary will be determined by type of material, monitoring, and the work activities required.

Warm Zone or Contamination Reduction (CRZ) or Decontamination Zone. This is a transition zone. Decontamination takes place in the warm zone in a designated area called the *contamination reduction corridor* (CRC). The degree of contamination increases as the line moves closer to the hot zone. Personal protective equipment is usually one level lower than that used in the hot zone. Depending on the material, however, the same level of PPE as used in the hot zone may be needed.

Support Zone (Clean or Cold Zone). This is the zone free of contamination. In the clean zone are administrative and other support functions that keep the warm and hot zones running smoothly. No PPE should be needed in this area.

9.2 DECONTAMINATION PLAN

A decon program must be developed as part of the site safety plan; it must be in effect before personnel or equipment enter areas where the potential for exposure to hazardous substances exists. When site conditions change, the plan must be modified to accommodate the new conditions.

The factors that must be considered in the design of the decon plan include, but are not limited to, the following:

Types of Contamination Different methods of removal are necessary for different substances and different levels of contamination. Contamination can be liquid or particulate; these differing states may require different methods of removal.

Contamination of personal protective equipment (PPE) can be categorized as surface contamination or contamination capable of permeating into the material. Surface contaminants are usually easier to detect and remove than contaminants that have permeated the PPE. Contaminants that permeate a material and are not removed by decontamination may be gradually released to surfaces where they may cause subsequent exposures.

Five major factors affect the extent of permeation:

- *Contact Time.* The longer a contaminant is in contact with an object, the greater the probability and extent of permeation. Therefore, minimizing contact time is one of the most important objectives of a decon program.
- *Concentration.* All molecules flow (diffuse) from areas of high concentration to areas of low concentration. As the concentration of contaminants increase, the potential for permeation of PPE increases.
- *Temperature.* As the temperature increases, the permeation rate of contaminant molecules increases.
- *Size of Contaminant Molecules and Pore Size of PPE.* Permeation increases as the size of the contaminant molecule decreases and as the pore size of the PPE increases.
- *Physical State of the Contamination.* As a rule, gases and low-viscosity liquids tend to permeate more readily than high-viscosity liquids or solids.

Any openings in PPE, such as cuts, punctures, inadequate zippers, or separations between two garments (e.g., pant legs and boots, cuffs, and gloves), may result in exposure to the wearer. Therefore, prior to each use, workers should inspect PPE to ensure that no such openings exist.

Common Decontamination Methods All personnel, clothing, equipment, and samples leaving the contaminated area of a site (hot zone) must be decontaminated to remove any hazardous chemicals or agents that may be adhering to or may have permeated their surfaces.

Physical Decontamination. The removal of gross contamination can be achieved by physical methods such as wiping, scraping, scrubbing, blowing, rinsing with flowing or pressurized water, or volatilization with steam.

The most common physical methods to decontaminate clothing and equipment involve the use of brushes, scrapers, or sponges. These methods are used in conjunction with detergent and water followed by water rinses to remove adhering contaminants. Figures 9.2 and 9.3 show decon personnel scrubbing boots with long-handled brushes and rinsing with hand sprayers to remove gross contamination.

FIGURE 9.2 ■ Decon worker scrubbing boots with a long-handled brush.

FIGURE 9.3 ■ Worker using a hand sprayer to decon workers.

Physical methods involving high-pressure water and steam should be used only if necessary and with caution because they can spread contamination and cause burns.

Chemical Decontamination. Physical removal of gross contamination should be followed by a wash/rinse process using cleaning solutions. These cleaning solutions normally utilize one or more of the following methods:

1. *Solubilization (Non-Water-Soluble Contaminants)*
   a. The chemical solvents chosen must be chemically compatible with the contaminants being dissolved and the equipment being cleaned. This is particularly important when decontaminating chemical protective clothing (CPC) made of organic materials that could be damaged or dissolved by organic solvents. In addition, care must be taken in selecting, using, and disposing of any organic solvents that may be flammable or potentially toxic. Organic solvents include alcohols, ethers, ketones, aromatics, straight-chain alkanes, and so on.
   b. Halogenated solvents such as carbon tetrachloride, methylene chloride, and so on, are incompatible with PPE and are toxic. They should be used only in extreme cases where other cleaning agents will not remove the contaminants.
2. *Dispersion (Water and Non-Water-Soluble Contaminants)*. Surfactants, detergents, and wetting agents are cleaning additives that augment physical cleaning methods by reducing the adhesion forces between contaminants and the surfaces being cleaned and by preventing the redeposition of the contaminants. Household detergents are among the most common surfactants. Some detergents can be used with organic solvents to improve the dissolving and dispersal of contaminants into the solvent.
3. *Rinsing (Water and Non-Water-Soluble Contaminants)*. Rinsing removes contaminants through dilution, physical attraction, and solubilization.

Many factors, such as cost, availability, and ease of implementation, influence the selection of a decon method. From a health and safety standpoint, two key questions must be addressed:

- Is the decontamination method effective for the specific substances present?
- Does the method itself pose any health or safety hazards?

Effectiveness Testing Decon methods vary in their effectiveness for removing different substances. The effectiveness of any decon method should be assessed at the beginning of the program and periodically throughout the lifetime of the program. If contaminated materials are not being removed or are penetrating protective clothing, the decon method must be revised. The following methods may be useful in assessing the effectiveness of the decontamination.

1. *Visual Testing.* No reliable test will immediately determine how effective decontamination is. In some cases, effectiveness can be estimated by visual examination.
   a. *Natural Light.* Discoloration, stains, corrosive effects, visible dirt, or alterations in clothing fabric may indicate that contaminants have not been removed. However, not all contaminants leave visible traces, and many of them can permeate clothing without being easily observed.
   b. *Ultraviolet Light.* Certain contaminants, such as polycyclic aromatic hydrocarbons, which are common in many refined oils and solvent wastes, fluoresce and can be visually detected when exposed to ultraviolet light. Ultraviolet light can be used to observe contamination of the skin, clothing and equipment.
2. *Swipe Testing.* Swipe testing provides after-the-fact information on the effectiveness of decontamination. In this procedure, a cloth or paper patch (swipe) is wiped over the surface of the potentially contaminated object and analyzed in the laboratory. Both the inner and outer surfaces should be tested. Skin may also be tested using swipe samples.
3. *Cleaning Solution Analysis.* Another method for testing the effectiveness of your decontamination procedures is to analyze for contaminants left in the cleaning solutions. Elevated levels of contaminants in the final rinse solution may suggest that additional cleaning and rinsing are needed.
4. *Permeation Testing.* Permeation testing for the presence of chemical contaminants in the fabric or material of the PPE requires that pieces of the PPE be sent to the laboratory for analysis.

9.2.1 Decontamination Facilities Design

How Long Will Work Continue on the Site? A long-term cleanup operation makes permanent, well-equipped decontamination stations feasible. Brief sampling or cleanup activities, which may move from spot to spot, require portable equipment. A "decon kit" would be appropriate, or, for companies with many workers, a mobile self-contained facility such as a trailer can be used.

How Much Space is Available at the Appropriate Location on the Site? The arrangement of decontamination stations will depend on the size, shape, and topography of the space available adjacent to the exclusion or hot zone.

What Utilities Will Be Available on the Site? Some means of obtaining or generating electricity will be required. If clean water is not available, it will have to be brought in. Some means of producing hot water or steam may be desired.

What Other Facilities Will Be Needed in the Decontamination Zone? Restroom facilities, an air tank change area, worker rest stations, and an emergency medical station may be required. Facilities for heavy equipment decontamination may be needed.

9.2.2 Decontamination Procedures

Decontamination procedures must provide an organized process by which levels of contamination are reduced. The decontamination process should consist of a series of procedures performed in a specific sequence. For example, outer, more contaminated items (e.g., outer gloves and boot covers) should be decontaminated and removed first. Decontamination and removal of inner, less contaminated items (e.g., jacket and pants) should occur next, followed by decontamination of the least contaminated items (air tanks, facepiece, and finally the inner gloves). The sequence of stations is called the *decontamination line* as shown in Figure 9.4.

Separate decon lines may be required if incompatible wastes are being handled at different work locations in the exclusion zone. Entry and exit points should be conspicuously marked, and the entry to the decontamination zone from the exclusion zone should be separate from the entry to the exclusion zone from the decontamination zone.

Decontamination lines are site-specific because they are dependent upon the types of contamination and the type of work activities on site.

Examples of decontamination line layouts and procedures for personnel wearing various levels of protection are show in Figures 9.5 to 9.9.

9.2.3 Recommended Equipment for the Decontamination

In selecting decontamination equipment, consider whether the equipment itself can be decontaminated for reuse, or whether it can be easily

FIGURE 9.4 ■ Decontamination line setup during a training session.

disposed of. The recommended types of equipment for the decontamination of personnel, personal protective clothing, and equipment are listed below:

- Drop cloths of visqueen or other suitable plastic material on which heavily contaminated equipment and outer protective clothing may be deposited
- Collection containers, such as drums or suitably lined trash cans, for storing disposable clothing and heavily contaminated personal protective clothing or equipment that must be discarded
- Lined boxes with absorbents for wiping or rinsing off gross contaminants and liquid contaminants
- Large galvanized steel or plastic tubs, or children's wading pools, to hold wash and rinse solutions; these should be large enough for a worker with booted feet to stand in and should have either no drain or a drain connected to a collection tank or appropriate treatment system
- Wash solutions selected to wash off and reduce the hazards associated with the specific contaminants
- Rinse sollutions selected to remove contaminants and contaminated wash solutions

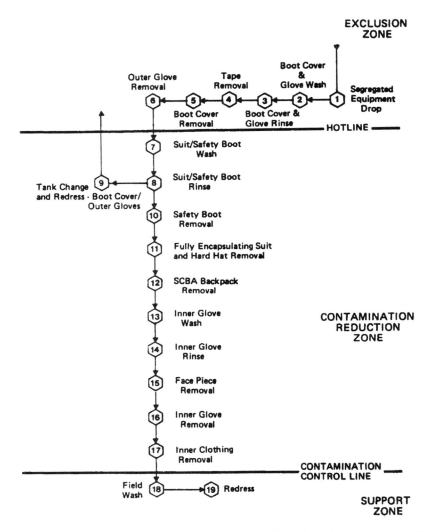

FIGURE 9.5 ■ **Maximum decontamination line layout—Level A protection.** *Source:* NIOSH, *Safety and Health Guidance Manual for Hazardous Waste Site Activities*, 1985.

- Long-handled, soft-bristled brushes to help wash and rinse contaminants
- Paper or cloth towels for drying protective clothing and equipment
- Lockers and cabinets for the storage of decontaminated clothing and equipment

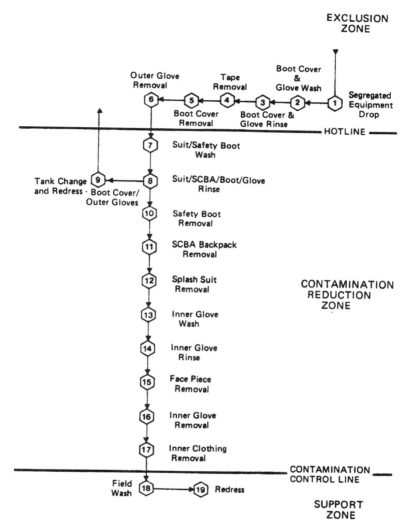

FIGURE 9.6 ■ **Maximum decontamination line layout—Level B protection.** Source: NIOSH, Safety and Health Guidance Manual for Hazardous Waste Site Activities, 1985.

- Metal or plastic cans or drums for the storage of contaminated wash or rinse solutions
- Plastic sheeting, sealed pads with drains, or other appropriate methods for containing and collecting contaminated wash and rinse solutions spilled during decontamination

302 ■ DECONTAMINATION

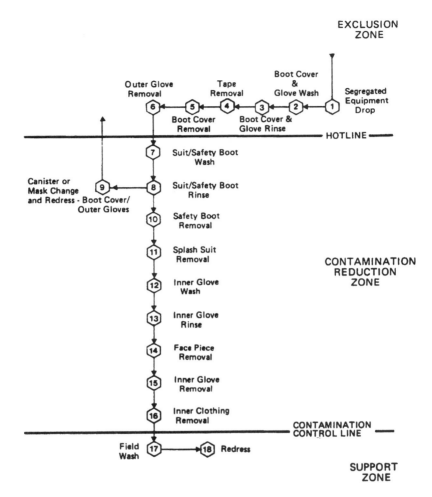

FIGURE 9.7 ■ **Maximum decontamination line layout—Level C protection.** *Source:* NIOSH, *Safety and Health Guidance Manual for Hazardous Waste Site Activities,* 1985.

- Shower facilities for full body wash or, at a minimum, personal wash sinks (with drains connected to collection tanks)
- Soap or wash solutions, wash cloths, and towels for personnel
- Lockers or closets for clean clothing and personal item storage

Other types of equipment not listed here may be appropriate in certain situations, and this list should not be considered all-inclusive.

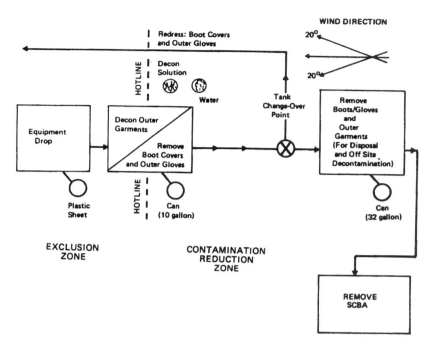

FIGURE 9.8 ■ **Minimum decontamination line layout—Levels A and B protection.** *Source*: NIOSH, *Safety and Health Guidance Manual for Hazardous Waste Site Activities*, 1985.

The Decontamination Line Entry and exit points to the decontamination line should be clearly marked. Workers entering the exclusion zone must pass through the decontamination reduction corridor by a route that does not permit them to be contaminated by workers in the decon line. If workers from different activities or parts of the exclusion zone require different decon solutions or procedures, more than one decon line must be arranged and clearly marked.

Special areas of consideration for inclusion in a decontamination line, if appropriate, are

- Tank change station, arranged at a point in the line where removal and replacement of the air tank does not permit contamination of the site worker or the decon helper
- Rest station beyond the point in the decon line where the necessary minimum decontamination has taken place

304 ■ DECONTAMINATION

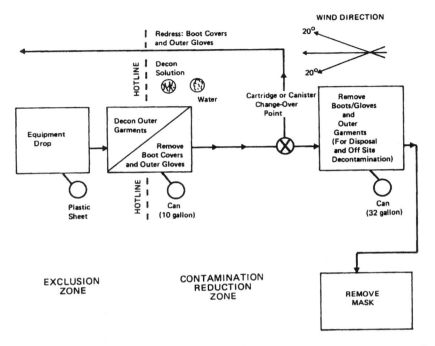

FIGURE 9.9 ■ **Minimum decontamination line layout—Level C protection.** *Source*: NIOSH, *Safety and Health Guidance Manual for Hazardous Waste Site Activities*, 1985.

■ Emergency medical station, together with an emergency decontamination plan

Protective clothing should be stored in such a way that air can circulate freely around the outside and inside of the garment, aiding in evaporation of the remaining solutions. Clothing contaminated by permeating chemicals should be disposed of if contaminants cannot be easily removed.

Decontamination of Tools Portable monitoring and sampling equipment may be partially protected from gross contamination by encasing the equipment in plastic. This practice will also reduce the necessary decontamination procedures.

In general, the decontamination of samplers, shovels, bung wrenches, and other tools is difficult and likely to be less than completely effective. As with chemical protective clothing, physical removal of liquids, residues, mud, and other thick materials is an important first step. Most tools can take stronger decon methods, such as chipping and steam jetting, than those used on personal protective clothing. Strong chemical solutions may also be used on equipment when appropriate, as long as the personnel using them are protected. However, care must be taken to avoid using strong acids, or caustics, or other harsh chemicals on sensitive metals and gasket components of the equipment being cleaned. Figure 9.10 shows a worker cleaning hand tools by vigorously scrubbing with detergent and water. All hand tools should be cleaned after each use and before storage.

Decontamination of Vehicles and Heavy Equipment Daily decontamination of heavy equipment on-site is impractical, and because most of this equipment is metal and relatively impermeable to most contamination, only partial decontamination is necessary. However, gross contamination should be removed daily to prevent buildup. Permeable surfaces such as tires, seats, and the inside of the operator's compartment should be completely decontaminated every day. All air vents and ducts should be flushed

FIGURE 9.10 ■ A decon worker dressed in Level C protection vigorously scrubbing hand tools with detergent and water.

with clean air as part of the decon process, and all heavy equipment having a grossly contaminated exterior should remain parked in the exclusion zone when not in use.

Vehicles hauling hazardous waste away from the contaminated areas must be thoroughly decontaminated before leaving the contamination reduction zone.

Recommended Equipment for Vehicle and Heavy Equipment Decontamination

- Pads or other containment structures for the collection of contaminated wash and rinse solutions. These containments should consist of storage tanks or appropriate treatment systems for the temporary storage and/or treatment of the contaminated wash and rinse solutions.
- Pumps for the collection and transfer of contaminated wash and rinse solutions.
- Wash solutions selected to remove and reduce the hazards associated with the contamination.
- Rinse solutions selected to remove contaminants and contaminated wash solutions.
- Pressurized water and steam are also used for the washing and rinsing of particularly hard-to-reach and clean areas as shown in Figures 9.11 and 9.12.
- Enclosures or curtains to contain splashes from pressurized cleaning equipment.
- Long-handled brushes, rods, and shovels for dislodging contaminants and contaminated soil caught in the tires and the undercarriage of vehicles and equipment. Figures 9.13 and 9.14 show workers decontaminating heavy equipment.
- Brooms, brushes, and wash-and-rinse buckets for use in the decontamination of the operator areas in and around vehicles and equipment.
- Containers for storage and disposal of contaminated wash-and-rinse solutions, damaged or heavily contaminated tools, and equipment.

Disposal Methods   All equipment used for decontamination must be disposed of properly. Buckets, brushes, clothing, tools, and other contaminated equipment should be collected, placed in containers, and labeled. Also,

FIGURE 9.11 ■ Decontamination of heavy equipment using a pressurized water sprayer.

all spent solutions should be collected and properly disposed of. Clothing that is not completely decontaminated should be placed in plastic bags or impermeable containers and disposed of properly.

Personnel Protection Decon workers may be exposed to hazardous substances from two sources:

- Workers, clothing, and equipment leaving the exclusion zone
- Decon solutions before, during, and after their use

Protective clothing and respirators that are appropriate to levels of specific contaminants with which they will come in contact must be selected

308 ■ DECONTAMINATION

FIGURE 9.12 ■ Pressurized steam is useful to clean contamination encrusted equipment and tools.

FIGURE 9.13 ■ A worker shown using pressurized water to decon a backhoe. *Source*: Courtesy of OHM Remediation Services Corp.

FIGURE 9.14 ■ Decon worker using a hose to dissolve and dislodge contamination from a vehicle hose fitting.

for decon personnel. Decon personnel who come in contact with workers and equipment leaving the exclusion zone will require more protection than decon workers assigned to stations at the end of the decon line. In general, decon workers should be dressed at the same level of protection or at one level below that of site workers, depending on their decon assignment.

All decon workers must be decontaminated before leaving the contamination reduction corridor, and their clothing, equipment, and solutions must be cleaned or disposed of properly.

Emergency Decontamination In addition to routine decontamination procedures, emergency decon procedures must be established. In an emergency, the primary concern is to prevent the loss of life or severe injury to site personnel. If immediate medical treatment is required to save a life, decontamination should be delayed until the victim is stabilized. If decon can be performed without interfering with essential lifesaving techniques or first aid or if a worker has been contaminated with an extremely toxic or corrosive material that could cause severe injury or loss of life, decontamination must be performed immediately.

If an emergency due to heat-related illness develops, protective clothing should be removed from the victim as soon as possible to reduce the heat stress.

During an emergency, provisions must also be made for protecting medical personnel and disposing of contaminated clothing and equipment.

9.3 SUMMARY

Decontamination is important to prevent the spread of hazardous materials beyond the scene of the incident. Proper procedures must be developed before an incident occurs. Precautions should be taken to prevent contamination of personnel and expensive equipment, such as monitors. During the size-up of an incident, work zones should be established according to the emergency response plan to control the spread of contaminants. There are three zones:

- The *hot zone* is the area where remediation activities are occurring. Only personnel in adequate PPE should be in this zone.
- The *warm zone or contamination reduction zone (CRZ)*, which surrounds the hot zone, is the area where decontamination occurs.
- The *cold zone or support zone* is the contamination-free area where support activities occur.

Methods to decontaminate personnel, PPE, and equipment will vary depending on the type of contaminant. Basic methods include

- Rinsing or dissolving
- Scraping, brushing, and wiping
- Evaporation, then rinsing
- Using surfactants, like soap
- Chemical disinfection
- Combinations of the above

The decontamination line is

- An organized series of procedures performed in a *specific* sequence
- Used to *reduce* levels of contamination on personnel, PPE, and equipment
- In operation until no contaminant is present

Each procedure is performed at a *separate station*. Stations are arranged in order of decreasing contamination, preferably in a straight line. Decontamination activities are located in the contamination reduction zone (CRZ).

All personnel working the decon line must be decontaminated before leaving the CRZ. All decon equipment must be properly decontaminated or disposed of properly.

REVIEW QUESTIONS

1. a. Design a minimum decontamination line to handle workers in Level B protection, remediating soil containing waste petroleum products.
   b. Design a maximum decontamination line to handle workers in Level B protection, remediating soil containing waste petroleum products.
   c. What additional decon stations did you add in the latter and why?
2. Why is the contamination reduction zone located as diagrammed in Figure 9-1?
3. How many decon lines would be required to handle the following three work crews and why?
   a. Work crew #1: remediating soil contaminated with metal plating bath chemicals (acids, chromium, nickel, cyanides, etc.).
   b. Work crew #2: sampling, transferring, and repacking drums containing numerous types of hydrocarbons (e.g., gasoline, heating oil, aromatic hydrocarbons, etc.).
   c. Work crew #3: excavating and removing underground storage tanks previously containing leaded gasoline.

   Diagram the line(s) you would recommend if you were in charge of the decon detail.

REFERENCES

EPA, *Standard Operating Safety Guides*, U.S. Government Printing Office, Washington, D.C., 1984.

Levine, S., and W. Martin, eds., *Protecting Personnel at Hazardous Waste Sites*, Butterworth, Stoneham, MA, 1985.

NIOSH/OSHA/USCG/EPA, *Safety and Health Guidance Manual for Hazardous Waste Site Activities*, U.S. Government Printing Office, Washington, D.C., 1985.

WORK PRACTICES

CHAPTER OBJECTIVES

When you have completed this chapter, you will

- Understand the importance of site characterization and analysis
- Know the reasons for the establishment of work zones on-site and what activities occur in each zone
- Understand the dangers of confined spaces and the procedures to follow when entering them
- Better understand the importance of emergency planning

10.1 PLANNING AND ORGANIZATION

Planning is the first and possibly the most critical step of hazardous waste site activities. By anticipating and taking steps to prevent or eliminate potential hazards to health and safety, workers at a waste site can proceed with minimum risk to themselves, the public, and the environment.

Planning for work site activities can be organized into three distinct phases:

1. The development of an overall *organizational structure*
2. The establishment of a comprehensive *work plan* that considers each specific phase of the operation
3. The development and implementation of a *site safety plan*

The organizational structure should identify the personnel needs for the overall operation, establish the chain of command, and specify the overall responsibilities of each employee.

The work plan should establish the objectives of site operations and the logistics and resources required to complete the specific elements of the project.

The site safety plan should determine the health and safety concerns for each phase of the operation and describe in detail the requirements, practices, and procedures as they relate to the workers, the public, and environmental protection.

Another important and sometimes overlooked aspect of planning is the coordination of site operations with existing response organizations. This coordination can often provide access to a variety of highly experienced personnel.

A national response organization was mandated by Congress under the National Contingency Plan to establish procedures and to coordinate response actions to releases of hazardous substances into the environment. This contingency plan established response teams composed of representatives of federal agencies and state and local governments. The EPA has designated individuals responsible for coordinating federal activities related to site cleanup.

Planning should be viewed as an ongoing process: the cleanup activities and site safety plan must be continuously modified to adapt to the changes encountered as site activities change and new hazards are revealed.

10.1.1 Initial Site Characterization and Analysis

Before any site operations begin, a thorough site characterization and analysis is required. The purpose of this activity is to identify the specific hazards present on-site so that appropriate protective measures can be taken.

Site characterization is usually carried out in stages. These are as follows:

Preliminary Evaluation. Before the team enters the site, as much information as possible should be collected. This information should include

- Hazards present
- Airborne toxic substances
- Combustible gases
- Oxygen-deficient conditions
- Ionizing radiation
- Site map showing the topographic features such as exact location, accessibility, and size of site
- Potential pathways of dispersion, wind direction, drainage, and so on
- Buildings, impoundments, tanks, pits, and ponds
- Description and expected duration of work activities expected to be performed on the site
- Identification (if possible) of areas on the site that require use of personal protective equipment

Perimeter Reconnaissance. After completion of the initial off-site information gathering, the perimeter of the site should be surveyed. This step will help in the completion of the preliminary site map and will determine the personal protective equipment needed for the initial site entry. While walking the site perimeter, the team should observe and record the location and condition of site buildings, tanks, drums, impoundments, distressed vegetation, surface soil and water characteristics, wind direction, and security concerns. Perimeter air and soil sampling for radiation, combustible gases, toxic substances, and other hazards is advisable during the survey, especially if the initial off-site characterization is incomplete.

10.1.2 Site Entry

Initial Entry Entry onto the site must be well planned and carefully executed. Personal protective equipment must be used, if required, to prevent worker exposures to the hazards expected to be on the site as a result of the preliminary evaluation.

If site hazards are not adequately identified during the preliminary evaluation, the workers on the initial entry team should be dressed at least in

Level B personal protective equipment and monitored for hazardous conditions as they enter the site.

The purpose of the on-site survey is to verify and supplement the information gathered from the off-site characterization. Prior to going on-site, workers should use the information gathered off-site to develop the required site safety plan. This plan must address the work to be accomplished, the procedures to follow, and the PPE that must be worn during the initial site entry.

The composition of the entry team depends on the site characteristics, but it should always consist of at least four persons: two workers who will enter the site and two outside support persons suited in PPE who are prepared to enter the site in case of emergency.

Upon entering the site the entry personnel should

- Monitor the air for IDLH and other conditions that might potentially expose workers to serious hazards (combustible gases, radiation, toxic substances, etc.)
- Complete the sampling required in the entry work plan, making sure to document all sampling procedures and locations and to label correctly all samples
- Review the site map and make the necessary changes to reflect the new information gained by the site entry
- Note the types and condition of waste containers and the presence of labels, tags, or other markings
- Identify any reactive, incompatible, flammable, or highly corrosive wastes
- Note the condition and contents of impoundments or other storage systems
- Determine the potential pathways for dispersion of the hazardous materials on-site:

    Air

    Biologic routes, such as animals

    Groundwater

    Surface water

- Note any indications of potential exposure hazards, such as dead animals or vegetation, airborne dust, and pools of liquids

- Note any safety hazards, considering the following:

    Condition of site structures

    Obstacles to entry and exit

    Terrain homogeneity

    Terrain stability

    Location of pits, wells, ponds, and lagoons

    Potential skin irritants such as poison ivy

Site Hazard Assessment Once the presence and concentrations of specific chemicals or classes of chemicals have been established, the health hazards associated with these chemicals must be determined. This can be accomplished by referring to standard reference sources for the permissible levels of exposure, lower and upper explosive limits, and so on.

Site Preparation After the initial site entry has been completed and the work plan and site map have been updated with the new information, the site must be prepared for the ensuing activities:

Time and effort must be spent in preparation of the site for cleanup activities. These preparations ensure that site operations go smoothly and that worker safety is protected. The following are routine preparations at typical cleanup sites.

- Roadway construction to provide access for personnel and heavy equipment
- Traffic flow patterns arranged to facilitate safe and efficient operations
- Elimination of physical hazards from the work site. These include

    Ignition sources in flammable hazard areas

    Exposed, underground, or overhead electrical wiring that might interfere with remediation activities

    Sharp or protruding edges, such as glass, nails, and torn metal, which can puncture personal protective clothing and equipment and inflict puncture wounds

    Debris, loose or protruding objects, slippery surfaces, unsecured objects, and holes

- Installation of skid-resistant strips on slippery surfaces
- Construction of operational pads for mobile facilities and temporary structures
- Construction of loading docks, processing and staging areas, and decontamination pads
- Provision of adequate illumination for work activities
- Use of ground fault interrupters on all temporary wiring

10.1.3 Site Control

Site Work Zones Another form of engineering controls is the establishment of site work zones. Such zones reduce the accidental spread of chemical contamination out of the contaminated area into the clean areas on- and off-site. These site work zones are delineated by the assessment of sampling and monitoring, and the visual survey of the site. A typical example of site work zones is shown in Figure 10.1.

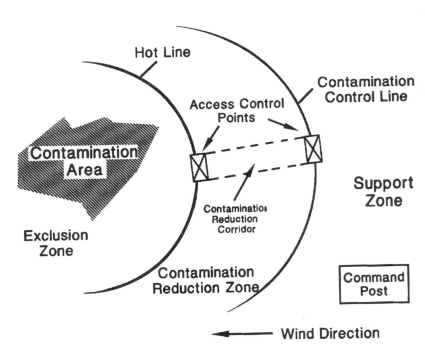

FIGURE 10.1 ■ **Diagram of site work zones.** *Source:* EPA, *Standard Operating Safety Guides,* 1984.

Hazardous waste sites are usually divided into three zones (but as many zones as needed can be used to protect the workers and the environment). The size and shape of each zone has to be based on conditions specific to each site. Figures 10.2 and 10.3 show the delineation of work zones for two specific sites. The zones described in the figures are as follows:

Exclusion Zone or Hot Zone. The area of maximum contamination is termed the *exclusion zone* or *hot zone*. The actual remediation activities occur in this zone, and the workers require the highest level of personal protective equipment. As the name implies, all personnel and equipment are to be excluded from this zone unless they have specific authorization and are properly protected in the prescribed level of protection.

The exclusion zone is laid out to include all areas of the site where contamination is known or expected to occur, as well as all areas where the

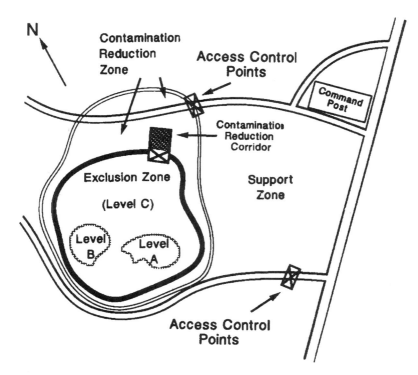

FIGURE 10.2 ■ **New Hampshire waste site.** *Source:* EPA, *Standard Operating Safety Guides,* 1984.

320 ■ WORK PRACTICES

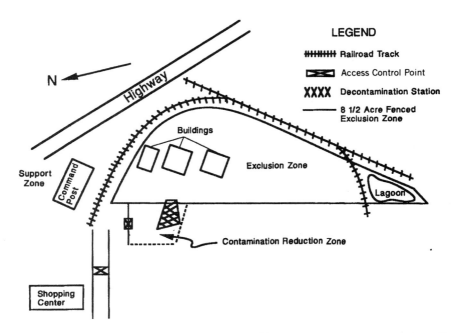

FIGURE 10.3 ■ **Lock Haven waste site.** *Source:* EPA, *Standard Operating Safety Guides,* 1984.

processing of hazardous wastes is planned. Figures 10.4 to 10.8 show different activities that do occur in exclusion zones.

This zone is surrounded by a boundary called the *hotline*. This boundary must be clearly marked, and often it is a physical barrier such as earthen berms, ditches, chain link fences or snow fences. Entrance into and out of the exclusion or hot zone must be regulated through access points. The corridor (termed the *contamination reduction corridor*) exiting the hot zone through these access points leads to the surrounding buffer zone or contamination reduction zone.

Contamination Reduction Zone. The area immediately adjacent to the hot zone is termed the *contamination reduction zone* (CRZ) or decontamination zone. This zone is a transition area between the contaminated area and the clean area or *support zone*. The CRZ is designed to reduce the probability that the clean support zone will become contaminated or affected by other hazards. The CRZ's location between the exclusion and sup-

10.1 PLANNING AND ORGANIZATION ■ **321**

FIGURE 10.4 ■ Inspection and sampling of drums.

FIGURE 10.5 ■ Inventorying of drums for shipment.

FIGURE 10.6 ■ Soil surface grading after remediation.

FIGURE 10.7 ■ The use of heavy equipment to remove contaminated soil and debris.

FIGURE 10.8 ■ The various types of heavy equipment used in the remediation process.

port zones, along with the decontamination of equipment, limits the physical transfer of hazardous chemicals into the clean areas. The outer boundary of the CRZ is called the *contamination control line*. This is the barrier through which no contamination should pass.

Decontamination procedures take place in a designated area within the CRZ called the *contamination reduction corridor* (CRC). The CRC controls access into and out of the exclusion zone and confines decontamination to a limited area. The size of the corridor depends on the number of stations in the decontamination procedure and the space available. The decontamination procedures begin as workers pass through the hotline and are completed prior to or at the contamination control line.

The level of contamination in the CRC decreases as one moves from the hotline to the contamination control line. The contamination control line is the boundary between the CRZ and the support zone.

Support Zone. This zone is the outermost part of the site; it is considered the noncontaminated or clean area. The support zone is the location of administrative and other support functions such as the staging area for support equipment, the command post, supply trailer, food wagon, and so on. The support zone and the rest of the site are secure areas bounded by fences or security controls. Traffic is restricted to authorized response personnel. Normal work clothes are appropriate within this zone.

The location of the command post and other support facilities in the support zone depends on a number of factors, including

- *Accessibility:* Topography, open space available, location of highways, railroad tracks, or other limitations.
- *Wind direction:* Preferably the support facilities and personnel should be located upwind of the exclusion zone. However, shifts in wind direction or other conditions may be such that an ideal location based on wind direction alone does not exist.
- *Resources:* Adequate roads, power lines, water, and shelter.

Support Zone Activities. The support zone is the location of many managerial and other support activities. These include the following:

Command Post
- Supervision of site operation and site workers
- Maintenance of communications, including emergency response and accident communications
- Record-keeping
- Public relations and media support personnel
- Monitoring of work schedules and weather reports
- Maintenance of site security
- Provision of sanitary facilities

Medical Station
- First-aid administration
- Medical emergency response
- Medical monitoring activities
- Sanitary facilities

Equipment and Supply
- Distribution center for equipment and clothing
- Maintenance and repair facilities for equipment and vehicles
- Receiving center for incoming equipment and expendable supplies

Administration
- Sample shipments
- Home–office interface

- Maintenance of emergency information such as telephone numbers, evacuation route maps, and vehicle keys
- Coordination center for transportation, disposal site, and appropriate federal, state, and local regulatory agencies

Field Laboratory

- Coordination and processing of environmental and hazardous waste samples
- Packaging and/or shipment of materials for off-site laboratory analysis
- Maintenance and storage of shipping papers, chain-of-custody files, laboratory notebooks

Figures 10.9 to 10.12 show several field laboratories and the routine testing and activities that occur on site.

Communication Systems Communication systems should be established as early as possible. Both internal and external systems may be necessary to ensure safe and efficient operations among site personnel.

FIGURE 10.9 ■ Field laboratory setup. Several workers dressed in Level C garb with face shields to provide protection from potential splashing during sample testing.

326 ■ WORK PRACTICES

FIGURE 10.10 ■ Closeup view of a technician using a Hnu to test samples for volatile organic compounds.

FIGURE 10.11 ■ Workers testing liquids for pH.

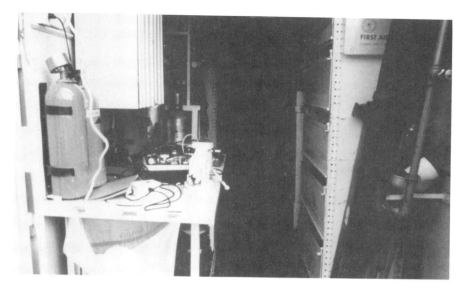

FIGURE 10.12 ■ Trailer converted to a mobil field laboratory.

Internal communication systems are used to

- Alert team members to emergencies
- Inform workers of changing site conditions, weather changes, safety information, and so on
- Communicate changes in work assignments, project changes, and so on
- Maintain site control

Verbal communication on-site can be impeded by background noise and the use of personal protective equipment. Respirator use can muffle sound and make communication difficult, and the use of protective hoods can also impair hearing. To improve site communication, workers can use hand signals and/or battery-operated radio devices. All electrically operated communication equipment used in potentially explosive atmospheres must be intrinsically safe.

External communication systems (between on-site and off-site personnel) are used to

- Coordinate emergency response
- Report to management

- Maintain contact with essential off-site personnel
- Contact regulatory agencies if necessary

The primary means of external communication are telephone and radio. With the growth and extensive use of cellular phones, this equipment has become the primary device for off-site communication.

Site Security Site security is necessary to

- Prevent the exposure of unauthorized, unprotected people to the site hazards
- Prevent the unauthorized dumping of hazardous wastes on the site
- Prevent vandalism and theft
- Avoid interference with safe working procedures

To maintain security, workers must

- Maintain security in the support zone and at access control points
- Establish an identification system to identify authorized personnel and verify their approved activities
- Erect a physical barrier or fence around the site if feasible, or post signs around the perimeter
- Assign security personnel to patrol and guard the site perimeter and entry and exit points
- Approve all site visitors, and provide guides if necessary
- Secure or lock all equipment when not in use

Figures 10.13 to 10.15 show several different security fences.

10.2 WORK PRACTICES

10.2.1 Confined Space Entry

On January 14, 1993, OSHA published its final rule 29 CFR 1910.146. This new regulation took effect on April 14, 1993. The new standard covers general industry and those confined spaces not covered by existing OSHA rules.

Confined space is a space that

- Is large enough and so configured that an employee can bodily enter and perform assigned work
- Has limited or restricted means of entry or exit, as with tanks, vessels, trenches, vaults, pits, bins, and hoppers

10.2 WORK PRACTICES ■ **329**

FIGURE 10.13 ■ Perimeter fences and guard to maintain site security by controlling access.

FIGURE 10.14 ■ Perimeter fence with barbed wire.

FIGURE 10.15 ■ Small portion of a perimeter fence stretching several miles around a site.

- Is not designed for continuous employee occupancy
- Has unfavorable natural ventilation

Limited Openings for Entry and Exit. Confined space openings are limited primarily by size or location. Openings are usually small, perhaps as small as 18 inches in diameter, and difficult to move through easily. Small openings make very difficult the passage of needed equipment in or out of the spaces, especially protective equipment such as respirators needed for entry into spaces with hazardous atmospheres, or lifesaving equipment when rescue is needed. However, in some cases, openings may be very large. Examples are open-topped spaces such as pits, trenches, and ships' holds. Access to open-topped spaces may require the use of ladders, hoists, or other devices, and escape from such areas may be very difficult in emergency situations.

Unfavorable Natural Ventilation. Because design features may not permit air to move freely in and out of confined spaces, the atmosphere inside a confined space can be very different from the atmosphere outside. Deadly or combustible gases may be trapped inside, particularly if the space is used to store or process chemicals or organic substances that may decompose. There may not be enough oxygen inside the confined space to support life, or the air could be so oxygen-rich that it is likely to increase the chance of fire or explosion if a source of ignition is present.

Not Designed for Continuous Worker Occupancy. Most confined spaces are not designed for routine worker entry. They are designed to store a product, enclose materials and processes, or transport products or substances. Therefore, occasional worker entry for inspection, maintenance, repair, cleanup, or similar tasks is often difficult and dangerous due to chemical or physical hazards within the space.

A confined space found on-site may have a combination of the preceding three characteristics, which complicate work performed in and around these spaces as well as rescue operations conducted during emergencies. If the survey of the site or facility identifies one or more of these spaces with the characteristics listed above, workers should follow the confined space entry procedures described in the following sections.

10.2.2 Permit-Required Confined Space Entry Procedures

The first task is to survey the work site to determine whether any permit-required confined spaces (spaces that meet the definition described above) exist. The next step is to determine if personnel must enter these spaces. If entry is required, the employer or the designated site or project manager must develop a written confined space program. Regardless of whether workers will enter permit-required spaces, measures must be taken to prevent entry by unauthorized people. This can be accomplished by posting warning signs.

The written confined space entry program, developed and/or implemented by the site manager, must contain the following 10 items:

1. *Hazard Identification.* The hazards presented by the permit-required confined spaces must be identified and evaluated prior to entry.
2. *Hazard Control.* Procedures, measures, and practices must be developed and implemented to eliminate or control confined space hazards. These procedures must include provisions for
   a. Implementing measures to prevent unauthorized entry
   b. Specifying acceptable entry conditions

c. Isolating the permit space and controlling hazardous energy (lock out/tag out)
   d. Purging, inerting, flushing, or ventilating the permit space as necessary to eliminate or control atmospheric hazards
   e. Providing barriers to protect authorized entrants
   f. Verifying that conditions remain acceptable through the duration of an authorized entry
3. *Specialized Equipment.* The site manager or the site safety officer must provide specialized equipment that may be needed to perform the permit-required confined space safety testing. Examples of such equipment may include air sampling instruments, ventilation equipment, personal protective equipment, lighting, communication equipment, barriers, and rescue equipment. The site safety officer must also verify that these items are being used properly and are maintained as required by the manufacturer. Figure 10.16 shows a worker using a combustible gas indicator and oxygen meter to identify potential hazards, and Figure

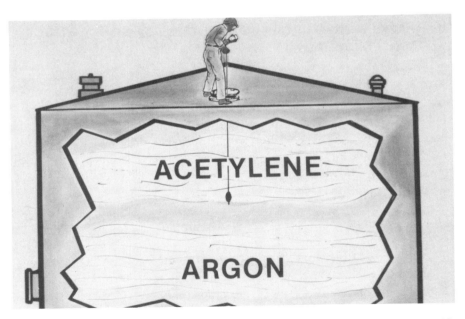

FIGURE 10.16 ▪ Worker testing a confined space for combustible gas and oxygen content prior to entry.

10.17 shows the entry team consisting of the entrant, the attendant, and the supervisor performing their required tasks.

4. *Testing and Monitoring.* Permit-required confined space conditions must be evaluated before and during entry operations. Pre-entry testing should be performed to the extent feasible before entry is authorized; additional monitoring may be required to determine if acceptable entry conditions are being maintained during the operations. Temporary ventilation systems are used when feasible. Figure 10.18 shows a ventilation system setup for a confined space. When testing the atmosphere, personnel are required to test for oxygen, then for combustible gases and vapors, and then for toxic gases and vapors.

5. *Attendants or Standby Persons.* At least one attendant outside the permit space is required during an authorized entry. This attendant must have the means and procedures to enable him/her to respond to an emergency situation in the space being monitored.

FIGURE 10.17 ■ Entry team with the required equipment needed for each member during entry. The team members are in both voice and visual contact continuously, and the entrant is attached to a retrieval system.

334 ■ WORK PRACTICES

FIGURE 10.18 ■ A temporary ventilation system consisting of a blower properly sized to provide adequate ventilation and enough duct work to reach to approximately 30 inches above the confined space floor or surface of the liquid in the space.

6. *Worker Designation.* Workers who play an active role in the entry, such as authorized entrants, attendants, and entry supervisors, must be specifically designated and trained in conformance with their anticipated duties.
7. *Rescue Procedures.* The site safety officer must develop and implement procedures for summoning rescue and emergency services, for rescuing entrants from permit-required spaces, and for preventing unauthorized personnel from attempting rescue.
8. *Permit System.* A written system for preparing, issuing, implementing, and canceling entry permits must be developed. This system must include provisions for closing the permit-required space and returning it to service after work is completed.
9. *Contract Employees.* The site manager or safety officer must coordinate entry operations when employees of more than one company are working simultaneously in a permit space.

10. *Permit and Program Review.* The permit-required entry operations should be periodically reviewed, and any problems encountered on the entry permit should be noted. Canceled permits should be reviewed at least once a year to determine if there is a need to modify existing procedures to ensure continued employee protection.

Permit System Before entry is authorized, the entry supervisor or site safety officer must complete the entry permit [see 29 CFR 1910.146(f)]. The entry supervisor must sign the permit and make it available to workers by posting it at the entrance to the permit space. The entry supervisor should cancel the permit when work in the space is completed or when an unexpected hazard exists inside or outside the permit space. All permits should be retained for one year for review purposes.

Entry Permit The standard requires that, at a minimum, entry permits include the following items:

- Identification of the space to be entered
- Purpose of entry
- Date and authorized duration of the permit
- Description of the hazards of the space
- Measures to be taken to isolate and manage the hazards of the space—that is, lock-out or tag-out of equipment, procedures for purging, inerting, ventilating, and flushing
- Acceptable entry conditions
- Results of initial and periodic air monitoring, including the name or initials of the tester and the time when the sampling was performed
- Communication procedures and equipment to maintain contact during entry
- Identity of authorized entrants and attendants
- Special equipment and procedures, including personal protective equipment and alarm systems
- Any additional permits such as those required for hot work
- Names and telephone numbers of the rescue and emergency services that can be summoned and the means for summoning them
- Entry supervisor's signature

- Any other pertinent information necessary to ensure worker safety in the particular confined space

Training Before workers are assigned confined space entry duties, they must receive the proper training to work in permit spaces. This training must provide workers with the knowledge and skills needed to perform safely their work in and around confined spaces.

The initial training program must make entrants, attendants and entry supervisors proficient in their required duties. Additional refresher training must be provided in any of the following situations:

- When there is a change in the workers' assigned duties
- When there is a change in permit-required confined space operations that present a hazard for which workers have not been trained
- When the site safety officer believes there are deviations from the permit-required confined space entry procedures or inadequacies in the workers' knowledge

Upon completion of the training, the employer or safety officer is required to certify that the training has been provided. The documentation must include the employee's name, signatures or initials of the trainers, and the date of the training. Documentation must be retained in the employee's training file and be available for inspection by the employees, their authorized representatives, and compliance officers.

Rescue and Emergency Services The confined space standard divides rescue capability into two categories, nonentry rescue and rescue service.

Nonentry Rescue To facilitate nonentry rescue, the standard requires that a nonentry retrieval system consisting of a rescue harness and lifeline be attached to each authorized entrant going into a permit space as shown in Figures 10.19 and 10.20.

Wristlets may be used instead of a harness if the employer can demonstrate that the use of a harness is infeasible or creates a greater hazard. The only exception to this requirement is in situations where the retrieval equipment would not help the rescue effort or would increase the overall entry risk.

FIGURE 10.19 ■ Worker donning a rescue harnesss.

One end of the retrieval line must be attached to the harness at the center of the entrant's back near shoulder level or above the entrant's head. The other end of the retrieval line must be attached to a mechanical device or fixed point outside the space to facilitate prompt response as shown in Figures 10.20 and 10.21.

If it is necessary to retrieve entrants from vertical permit-required confined spaces deeper than 5 feet, a mechanical lifting device must be available.

Rescue Services Rescue services may be provided by either on-site or off-site personnel. On-site rescue teams have the advantage of being immediately available and intimately familiar with the permit space. Because

FIGURE 10.20 ▪ Worker entering a confined space wearing a rescue harness attached to a retrieval system.

response time is critical in any emergency, an on-site team reduces response time to a minimum. The main disadvantage with an on-site team is the time and expense of initial and annual training as well as the cost of primary and backup rescue equipment.

At a minimum, the standard requires that rescue team members receive the same level of training as authorized entrants. They must also be trained in the proper use of personal protective and rescue equipment. Each member of the team must be trained in basic first aid and cardiopulmonary resuscitation (CPR), and at least one member must be currently certified in both.

FIGURE 10.21 ■ The rescue of an injured worker from a confined space using a retrieval system.

Rescue team members must also practice making permit space rescues at least once every 12 months. They should employ simulated rescue operations in which they remove dummies, mannequins, or actual persons from the actual permit space or from representative permit spaces, provided that the size, configuration, and degree of accessibility approximate the spaces from which rescue may be performed.

If an employer chooses to use an off-site rescue service, such as another employer's workers (contract rescue service) or the local municipal rescue services, the employer must do two things:

- Inform the rescue service of the hazards that may be encountered when responding to an incident involving the permit space.
- Provide the rescue service with access to all permit-required spaces from which rescue may be necessary, develop appropriate rescue procedures and conduct practice operations.

Additionally, because personnel who may become injured in a confined space may also have been exposed to or contaminated by hazardous materials, written information must be available at the space for these substances. This information must be provided to the medical facility treating the exposed employee.

10.2.3 Emergency Response Planning

Working at hazardous waste sites can involve handling of flammable, toxic materials present in bulk quantities under a variety of physical and environmental (heat, cold, rain, uneven topography) circumstances. The nature of this work makes emergencies a continual possibility, no matter how frequently they may actually occur. Emergencies must be planned for in advance to ensure a fast, appropriate response to a specific incident. However, at times, this planning seems impossible because each site is different and many different kinds of emergencies can occur in thousands of different combinations. Therefore, effective planning and training for four or five basic emergencies with the necessary equipment will allow workers to deal effectively with most emergencies.

Emergency Recognition and Prevention On a day-to-day basis, site personnel should be constantly alert for indications of potentially hazardous situations. Rapid recognition of potentially dangerous situations can avert an emergency. Better still is the advance planning of operations to anticipate and control hazards before they arise. Recognition of potential hazards should include answering the following questions:

- *Chemical Hazards.* Are MSDSs available for materials at the sites? Is sufficient PPE and respiratory protection available for personnel handling these materials? Are site operations to occur near areas where unprotected personnel reside or work? Are the materials especially reactive with water, air, or other chemicals on the site? What steps can be taken to limit this reactivity? Are air monitoring instruments available or required? Are there any ventilation intakes near the work site?
- *Physical Hazards.* Will uneven ground, slippery conditions, presence of pits or lagoons, and work at heights require specials procedures or equipment?
- *Environmental Hazards.* Will work occur at night? Will heat stress or cold stress be a factor? Could adverse weather conditions hamper operations?

- *Emergency Procedures.* Are special antidotes and other special medical procedures required in the event of exposure? Is special firefighting equipment required?
- *Training.* Are the workers and other personnel involved at this site experienced with site operations? Are the workers familiar with the handling procedures for the chemicals present on-site? Are there any new pieces of equipment on-site with which the working crew is unfamiliar? Are all workers familiar with the procedures to be followed in the event of an emergency?
- *Tasks to Be Performed.* Trenching and excavations may require sloping and shoring. Are structures close enough that undermining is a concern? Are buried utilities located in the areas of excavation?

Site Emergencies: Roles, Communications, and Evacuation Procedures

Many emergencies will require a coordinated effort from the entire crew to adequately cope with the problems at hand. The following describes the basic structure and organization that is required for an emergency response.

Roles, Responsibilities, and Lines of Authority In the event of an emergency, one person must be vested with the authority to direct actions and make decisions. This person is called the *on-scene commander* or *emergency response coordinator*. On construction projects, the site supervisor usually assumes this role with others reporting to him/her.

In an emergency such as an explosion or fire, all personnel on-site must be accounted for. This information should be communicated to the emergency response coordinator through some form of organized reporting structure. Most companies will use the same reporting mechanism in dealing with emergencies as with normal operations.

Alarm Signals A signaling system must be developed and implemented to alert personnel of an emergency. The requirement for these signals are as follows:

- Signals must be capable of being heard by all personnel as an indication of an emergency situation. On small projects, voice contact and hand signals may be sufficient.

- Multiacre sites and noise from equipment present problems in that usually no one signal will adequately work, but a combination should be used.
- A workable combination may be a blast on a hand-held horn followed by a radio transmission.
- Different combinations of signal blasts can be used, such as short–short or long–long to denote separate events—for example, medical emergency, severe weather, or fire. But the use of multiple signals will usually result in confusion. If two separate signals are required, a single series of short blasts and/or long blasts does work. Remember to keep it simple.
- Check if the facility has its own established signal sets and what they mean. Providing workers with information typed on cards and taped to the inside of their hard hats can help.

Site Evacuations—Refuges (Safety Stations) No single recommendation can be given for evacuation or safe distances because of the wide variety of hazardous substances and releases found at sites. For example, a 'small' ammonia leak may require an isolation of 150 feet, whereas a large leak may require the evacuation distance of 1 mile or more, depending on the wind direction.

Safe distances can only be determined at the time of the emergency, based on the combination of site and incident specific factors. Planning and practicing emergency response to a variety of scenarios will help familiarize site personnel with the standard operating procedures to follow in the event of different emergencies.

In the event of severe weather or uncontrolled or catastrophic accidents, a complete evacuation of the site may have to be declared until the incident is safely addressed. A refuge or designated gathering points must be selected in order to regroup and count heads. These refuges should also provide the necessary first aid for injured personnel and provide communications with the command post.

Emergency Telephone Numbers During an emergency, the last thing one wants to do is to locate telephone numbers from a telephone book. OSHA, under 29 CFR 1926.50(f), requires that telephone numbers of physicians, local hospitals, or ambulances be conspicuously posted. Common practice also requires posting of telephone numbers for the police and fire departments. Numbers of the local EPA or DNR/DEP representative along with the National Response Center may also be posted. Other points to consider are as follows:

- These lists should be positioned near each site telephone.
- If there is no site telephone, the list should still be posted and include the location of the nearest telepone.
- Call each of the numbers to make sure they are correct. It is very easy to transpose a phone number.
- Call each emergency agency to ensure that they will respond to your location. For example, projects located outside city limits may be referred to by the city name but may be in the jurisdiction of the county sheriff.

Offsite Response Parties Several off-site response parties may need to respond to the site in event of an emergency. The following points should be considered:

Police Police will usually respond to direct traffic or file reports if any ambulance or fire calls are made.

It may be a good idea to alert the local police and let them know about the project location. In most major cities, the police probably will not care. Small police departments may be very interested in the project.

Ambulance Service Not every city has a city-run ambulance/EMT squad. Sometimes this service is covered by a county organization or private companies. You will want to identify which one will respond to your site.

There may be multiple private ambulance companies. You should investigate to determine the best response time, the best-equipped units, and the best-trained people.

Contact the selected ambulance service and alert them to the project and contaminants. The last thing you want to happen is having an ambulance crew respond to the site and then refuse to assist an injured worker for fear of contamination. The service may want to visit the site and review operations.

Ambulances may want assurances that protective materials, such as (a) visqueen to line the vehicle's interior and (b) tyvek coveralls, booties, and gloves for responding personnel, will be available if the material is considered hazardous.

Hospitals   Hospitals must be selected to meet the following needs:

When working with especially toxic materials present in large quantites—for example, cyanides, anilines, pesticides and certain gases such as phosgene—the nearest hospital emergency room should be contacted. You will want to ask if they know the treatment for such chemicals and if antidotes, (cyanide, pesticides) are available. If there is a problem, contact the regional Poison Control Center.

Some hospitals in large cities are equipped with a designated decon area for chemically exposed patients. Usually this is referred to as having the capacity to deal with "chemical trauma cases." If hospitals do not have this capability, they may not want to deal with your personnel and may refuse treatment.

Not every injury needs to be treated in a hospital emergency room, which is usually the most expensive route of treatment. In some cases, the small clinics are open long hours and can take care of small cuts, burns, and regular illness problems (colds, flu, etc.).

Emergency Map   Not every medical emergency will require an ambulance to respond to the site, but an emergency map should be developed showing the route from the project site to the nearest hospital or clinic. An emergency map should be comprised of two parts: a diagrammatical sketch and written instructions. Be sure to include both the hospital's telephone number and that of the site. Determining the route to the hospital from the hotel also may be worthwhile.

First Aid Kits   Several common types of first aid equipment are found at hazardous waste sites. Two specific OSHA regulations define the requirements for first aid kits. These requirements are as follows:

- 29 CFR 1926.50(d)(1). First aid supplies approved by the consulting physician shall be easily accessible when required.
- 29 CFR 1926.50(d)(2). The first aid kit shall consist of materials approved by the consulting physician. Materials must be kept in a weatherproof container with individual sealed packages for each type item. The contents of the first aid kit should be checked weekly

by the employer before the kit is sent out on each job. The kit should be checked at least weekly on each job to ensure that the expended items are replaced.
- OSHA does not specifically dictate the contents of a kit. OSHA does not specify the ratio of first aid kits to workers. However, the Army Corps of Engineers Safety and Health Requirements Manual states that one small kit should be available for every six workers, or one large kit should be available for every 25 workers.
- 29 CFR 1926.50(e) states the following "Proper equipment for prompt transportation of the injured person to a physician or hospital, or a communication system for contacting necessary ambulance service, shall be provided." This standard reinforces the need for emergency contacts to be established. Depending on the work site and training of personnel, a stretcher is advisable. A portable oxygen unit and a back board are advisable if employees have proper training in their use.

First Aid Training. 29 CFR 1926.50(e) In the absence of an infirmary, clinic, hospital, or physician, who is reasonably accessible in terms of time and distance to the work site, a person who has a valid certificate in first aid training from the U.S. Bureau of Mines or the American Red Cross, or equivalent training verifiable by documantary evidence, must be available at the work site to render first aid. Good practice would indicate that at least one crew member should be trained in first aid or CPR.

Eyewash Equipment and Safety Showers OSHA recognizes the importance of suitable drenching facilities for personnel accidentally splashed with corrosive liquids. 29 CFR 1910.151(c) states the following: "Where the eyes or body of any person may be exposed to injurious corrosive materials, suitable facilities for quick drenching or flushing of the eyes and body shall be provided within the work area for immediate emergency use."

Eyewash Equipment Eyewash units shall be in accessible locations that require no more than 10 seconds to reach and should be within a travel distance no greater than 100 feet from the hazard.

- Eyewash equipment shall be capable of delivering to the eyes not less than 0.4 gallons per minute (1.5 liters) for 15 minutes.

- The unit should be between 33 and 45 inches from the floor.
- Eyewash location should be marked.
- Eyewash units should be checked at least weekly, and water should be changed monthly. Antibacterial/fungal agents can be purchased and added to the water. Regular potable water is satisfactory. Deionized water is not necessary.
- Units need to be protected from freezing. Special units are made with internal heaters, although these are expensive. Units can be successfully insulated and heat-taped to prevent freezing.
- Nozzles should be protected from airborne contaminants (dust, dirt).
- Workers should be trained in the correct use of the the equipment.

Safety Showers Emergency showers shall be in accessible locations that require no more than 10 seconds to reach and should be within a travel distance no greater than 100 feet from the hazard. With extremely corrosive materials, the shower facility should be closer. This standard was developed for people not wearing PPE.

The shower must be high enough for people to get underneath. It should have a spray diameter of 20 inches wide, 60 inches above the surface. It should deliver 30 gallons per minute. The control valve should remain on without use of operator's hands. Showers need to be checked weekly.

In remote areas, a portable shower could consist of a shower stand, a hose, an overpack drum of water, a portable generator, and an electric pump.

The goal is to wash the body for 15 minutes following chemical exposure.

Some provision may be necessary to capture used water.

Fire Protection OSHA regulates fire protection standards in 29 CFR Subpart 1926.150 and 29 CFR Subpart L 1910.155-165. In general, these standards require the following:
- A communication system for employees to contact the fire department.
- If portable fire extinguishers are provided, they shall be the proper type and size.
- Employees should not travel more than 75 feet (29 CFR 1910) or 100 feet (29 CFR 1926) to find an extinguisher.
- Portable extinguishers should be visually inspected monthly and professionally inspected annually.

- Personnel should be trained in the proper use of firefighting equipment.
- Portable fire extinguishers need to be secured, upright and identified.
- No smoking or open flames are permitted in areas used for fueling. Signs should be posted to that effect.

Spill Response Most operations on a hazardous waste site involve pumping, transferring and handling hazardous materials. No matter how carefully performed these operations will occasionally result in materials being spilled. Procedures and materials should be in place to allow a safe and efficient, subsequent cleanup.

> *Preoperational Inspections.* A careful inspection of transfer equipment—that is, hose connections and pumps—should be made to ensure that connections are tight, hoses are of proper length and in good condition, and pump/hose systems are properly sized. Inspect and sound any tank to determine the volume of material to be transferred. The transfer tankage (drums, bulk tank) should have sufficent volume to contain these fluids. Remember that pumping will sometimes cause frothing, which will increase fluid volume. Temperature differences between in-ground tank fluid and the ambient temperature may cause an increase in volume.
>
> *Spill Control Agents.* Spill sorbents should be available to control spilled materials. Spill sorbents selection should be based on the chemical properties of the tank material. The control agent for an organic solvent may not be appropriate for an acidic or basic material.
>
> *Transfer or Rendering Site Preparations.* The material transfer area or rendering site should be prepared to ensure proper confinement in event of a spill. This preparation may be as simple as lining an area with visqueen and berming with soil. Contractual requirements may specify a metal or concrete pad with proper sloping and sumps to be installed prior to removal.
>
> *Safety Equipment.* Appropriate safety equipment, which includes respiratory protection, chemical protective clothing, and air monitoring devices, should be available in the event of an accidental release.

Personnel must be properly trained in the use and limitations of these devices. Eyewashes and safety showers should be available and are required.

Emergency Notifications. In the unlikely or unfortunate event that chemical materials either leave the property, enter a sanitary sewer system, or enter a navigable waterway, emergency telephone numbers should be posted for proper notification.

Disposable Containers. If a spill occurs, suitable and sufficient containers should be available for disposal of the material and chemical control agents.

Accident Investigation and Follow-Up Accidents are undesired events that result in physical harm and/or property damage. Accidents usually result from contact with a source of energy above the threshold limit of the body or structure.

Accident investigations should gather sufficient information to portray accurately the circumstances of the accident and to prevent the problem or situation from recurring.

Numerous texts describe the protocol and techniques for conducting accident investigations. Each author has his/her own view of the proper way to perform an investigation; however, they do concur on several points:

1. There is no such thing as a minor accident, although some accidents result in only minor injuries or damages.
2. The goal is fact-finding, not fault-finding.

Investigations are most successful if conducted immediately. The longer the delay, the greater the possibility of obtaining incomplete or inaccurate facts, because physical locations change, memories fade, collective discussion begins to influence memories, and emotional shock clouds objective reasoning.

In a systematic accident investigation, the first task is to collect the facts. Facts to be recorded include time and location of the incident, the environmental conditions, the position of people, equipment, and material, and the relationships between these items. Witnesses along with personnel should be interviewed individually, helping to prevent collective discussions from influencing the facts.

With the facts collected, an analysis should be made to determine the causative factor of the accident. Rarely will all the facts be completely supportive of one another; at times they may be contradictory. This again could occur because memories of participants might be colored by the experience, thus their accounts may not match the physical evidence that is present.

The conclusion should summarize the fact collection and analysis phases of the investigation and indicate the cause of the accident. Multiple causes can sometimes be reached with a primary and secondary conclusion.

Finally, the most important section of an accident investigation is the recommendation on how to prevent the accident in the future.

Regulatory mandate for accident reporting to OSHA, 29 CFR 1904, requires most employers to maintain logs—that is, an OSHA 200 form noting accidents that require medical attention (other than simple first aid) or lost work days.

OSHA also requires that fatalities of one or more employees or hospitalizations of five or more employees be written or orally reported to the area director within 48 hours of occurrence.

REVIEW QUESTIONS

1. What is the major purpose of site characterization and analysis?
2. Why are sites divided into work zones?
3. What major factors determine the location of the support zone?
4. List two air monitoring instruments used before entry and while working in confined spaces. Why is each instrument important?
5. What work site information must be supplied to an ambulance service?
6. What information must the hospital receive to provide the appropriate care?

REFERENCES

EPA, *Standard Operating Safety Guides*, U.S. Government Printing Office, Washington, D.C., 1984.

Levine, S., and W. Martin, eds., *Protecting Personnel at Hazardous Waste Sites*, Butterworth, Stoneham, MA, 1985.

Pettit, T., and H. Linn, *A Guide to Safety in Confined Spaces*, Department of Health and Human Services (NIOSH) No. 87-113, U.S. Government Printing Office, Washington, D.C., 1987.

NIOSH, *Working in Confined Spaces, Criteria for a Recommended Standard*, NIOSH No. 80-106, U.S. Government Printing Office, Washington, D.C., 1979.

NIOSH/OSHA/USCG/EPA, *Safety and Health Guidance Manual for Hazardous Waste Site Activities*, U.S. Government Printing Office, Washington, D.C., 1985.

OSHA 29 CFR 1910.120, *Hazardous Waste Operations and Emergency Response (SARA)*, Washington, D.C., U.S. Government Printing Office, 1989.

OSHA 29 CFR 1910.146, *Permit—Required Confined Spaces for General Industry*, U.S. Government Printing Office, Washington, D.C., 1993.

Waxman, Michael F. ed., *Confined Space Entry* University of Wisconsin Workshop, November 9–10, 1994.

DEVELOPING A SITE SAFETY PLAN

CHAPTER OBJECTIVES

When you have completed this chapter you will

- Know what a site safety plan consists of
- Be able to read and understand your site plan

The development and implementation of a written site safety plan is mandated by the Superfund Amendments and Reauthorization Act of 1986 and is required by OSHA in the standard 29 CFR 1910.120.

The purpose of a site safety plan is to establish requirements for protecting the health and safety of site personnel during all activities at that site. OSHA requires that all site personnel abide by provisions of the plan.

The site safety plan must be prepared and periodically reviewed by qualified personnel to keep it current and technically correct for all hazardous substances known to, or expected to, exist at that specific site. The site safety plan will include much of the information included in previous chapters under initial site characterization, site entry, site control, confined space entry, and emergency response planning.

For long-term remedial activities at hazardous waste sites, safety plans should be developed simultaneously with the general work plans. Workers should become familiar with the plan before worker activities begin.

The plan must contain safety requirements for routine (but hazardous) activities and also for unexpected site emergencies. The major distinction between routine and emergency site safety planning is predictability: site personnel can predict, monitor, and evaluate routine activities, but site emergencies are unpredictable and may occur anytime.

11.1 GENERAL REQUIREMENTS

The site safety plan must contain the following items to comply with OSHA, 29 CFR 1910.120:

- A hazard assessment, based on a thorough site characterization for each phase of the activities involved
- The name of the designated site safety officer and alternates responsible for site safety, response operations, and protection of the public
- Description of the personal protective equipment or levels of protection to be worn by the site personnel
- Site map with delineated work areas or work zones
- Established procedures to control site access
- Description of decontamination procedures for personnel and equipment
- Established site emergency procedures to handle any anticipated site emergency; description of procedures for medical care of injuries and toxic exposures
- Description of medical surveillance and air monitoring requirements
- Training requirements and assignments
- Established confined entry procedures
- Established spill containment program
- Description of procedures for trenching and excavations
- Description of all sampling procedures

Modification to the Plan Site safety planning should be initiated prior to the beginning of on-site operations. An initial plan should be developed to cover initial site entry and modified as the site characterization progresses. The plan should be modified whenever

- Site conditions change
- Different phases of site operations are initiated
- Additional site information becomes available

Prior to site entry and as conditions change, all site personnel, including employee representatives, contractors, and subcontractors, must read the plan and be routinely briefed on the primary contractor's site safety plan or their own company's plan. The plan must be posted on-site to be available to all affected personnel.

11.2 SITE SAFETY PLAN

The following site safety plan (Generic Site Safety Plan, Appendix B of the *Occupational Safety and Health Guidence Manual for Hazardous Waste Site Activities*, NIOSH, OSHA, USCG, EPA, 1985) is a highly generalized example; it will require extensive modification to produce an acceptable *site-specific* safety and health plan. It should be used as a guide only, not as a standard.

GENERIC SITE SAFETY PLAN
(Suggested format for minimum site safety plan)

A. Site Description

Name of Hazardous Waste or Project Site _____

_____

Site Location _____

_____

Date _____

Hazards/Risks _____

Area/Location Affected _____

_____

Surrounding Population _____

Topography _____

Prevailing Weather Conditions _____

_____

_____

Additional Information _____

_____

_____

_____

B. **Entry Objectives** (Initial entry objectives—actions, tasks to be accomplished; i.e., identify contamination, determine levels of contamination, monitor conditions, take samples, etc.)

_____

_____

_____

_____

C. **On-Site Organization and Coordination** (The following personnel are designated to carry out the stated job functions on-site.)

Project Team Leader _____

Science/Expert Advisor _____

Site Safety Officer _____

Public Information Officer _____

Security Officer _____

Recordkeeper _____

Financial Officer _____

Field Team Leader _____

Field Team Members _____

_____

_____

_____

Federal Agency Representatives

(OSHA, EPA, etc.) _____

_____

_____

State Agency Representatives

(DNR, etc.) _____

_____

_____

Local Agency Representatives

(Fire Marshal, Police, etc.) _____

_____

_____

Contractor(s)

_____

_____

_____

All personnel arriving or departing the site should log in and out with the Recordkeeper. All activities on-site must be cleared through the project team leader.

D. On-Site Control

(Name of individual or agency) has been designated to coordinate access control and security on-site.

A safe perimeter has been established at (distance or description of controlled area) _____

_____

_____

No unauthorized person should be within this area.

The on-site command post and staging area have been established at _____

_____

_____

_____

The prevailing wind conditions are _____. This location is upwind from the exclusion or hot zone.

Control boundaries have been established, and the exclusion zone (the contaminated area), hotline, contamination reduction zone, and support zone (clean area) have been identified and designated as follows: <u>(describe boundaries and/or attach map of controlled area)</u>

_____

_____

These boundaries are identified by: <u>(marking of zones, i.e., snow fence—the hotline and the contamination control line; red boundary tape—the decontamination line or CRC; traffic cones—rest areas in support or decon zone, etc.)</u>

_____

_____

E. Hazard Evaluation

The following substances are known or suspected to be on site and the primary hazards of each are identified below.

Substance Involved (chemical name)	Concentrations (if known) (ppm, mg/l and/or %)	Primary Hazards (flammability, etc.)
Naphtha (in soils)	100–450 ppm	Flammable/toxic

11.2 SITE SAFETY PLAN ■ 357

spent sulfuric acid	2–17%	corrosive/reactive
_____	_____	_____
_____	_____	_____

The following additional hazards are expected on site: (i.e., slippery ground, uneven terrain, open pits, poison ivy, buried construction debris, etc.)

Hazardous substance information forms and/or MSDSs for the involved substances listed above and for chemicals brought to and used on site (i.e., decontamination solutions, degreasers, lubricants, etc.) have been completed and are attached.

F. Personal Protective Equipment

Based on the evaluation of potential hazards, the following levels of personal protection have been designated for the applicable work areas or tasks:

Locations	*Job Function*	*Level of Protection*
Exclusion Zone	_____	A B C D Other
	_____	A B C D Other
	_____	A B C D Other
	_____	A B C D Other
Contamination	_____	A B C D Other
Reduction Zone	_____	A B C D Other
	_____	A B C D Other
	_____	A B C D Other

Others

Specific protective equipment for each level of protection is as follows:

DEVELOPING A SITE SAFETY PLAN

Level A fully encapsulated, chemically resistant suit

chemically resistant boots

disposable inner gloves and disposable coveralls

pressure demand SCBA

_____

_____

Level B chemically resistant clothing or splash gear (hooded, long-sleeved coveralls or hooded jacket and pants and/or splash apron)

chemically resistant boots

pressure demand SCBA

disposable inner gloves

_____

_____

Level C chemically resistant clothing

full-facepiece air-purifying respirator

chemically resistant boots

disposable inner gloves

_____

_____

Level D recommended: coveralls

safety boots/shoes

safety glasses/chemical splash goggles

hard hat

_____

11.2 SITE SAFETY PLAN ■ 359

Other _____

_____

_____

The following protective clothing materials are required for the involved substances:

Substance	*Material*
(chemical name, e.g., gasoline, PCBs mineral spirits, etc.)	(material name, e.g., PVC, Viton, Saranex, tyvek, etc.)
_____	_____
_____	_____
_____	_____
_____	_____
_____	_____
_____	_____

If the use of air-purifying respirators is authorized, the _____ _____ is the appropriate air-purifying cartridge/canister for use with the following involved substance _____ _____ at concentrations _____ _____ ppm/mg/liter.

If the use of air-purifying respirators is authorized, the _____ _____ is the appropriate air-purifying cartridge/canister for use with the following involved substances _____ _____ at concentrations _____ _____ ppm/mg/liter.

■ DEVELOPING A SITE SAFETY PLAN

No changes to the specified levels of protection shall be made without the approval of the site safety officer and the project team leader.

G. On-Site Work Plans

Project team(s) consisting of _____ persons will perform the following tasks:

Field Team Leader (name) _____   _____(function)_____

_____

_____

Field Team #1 _____   _____

(work party)   _____

_____

_____

Field Team #2 _____   _____

_____

_____

_____

Rescue Team _____   _____

(if required)   _____

_____

_____

Decontamination _____   _____

Team   _____

_____

The field teams or work parties were briefed on the contents of this plan at _____

H. Communication Procedures

Channel _____ has been designated as the radio frequency for personnel in the exclusion zone. All other on-site communications will use channel _____ .

Personnel in the exclusion zone should remain in constant radio communication or within sight of the project team leader. Any failure of radio communication requires an evaluation of whether personnel should leave the exclusion zone.

(Horn blast, siren, etc.) is the emergency signal to indicate that all personnel should leave the exclusion zone.

The following hand signals will be used in case of failure of radio communications:

Signal	Indication
Hand gripping throat	Out of air, can't breathe
Grip buddy's (partner's) wrist or both hands around waist	Leave area immediately
Hands on top of head	Need assistance
Thumbs up	OK, I am all right, or I understand
Thumbs down	No, negative

Telephone communication to the command post should be established as soon as practicable. The phone number is _____ .

I. Decontamination Procedures

Personnel and equipment leaving the exclusion zone shall be thoroughly decontaminated. the standard level A, B or C decontamination protocol shall be used with the following decontamination stations:

(1) _____ (2) _____

(3) _____ (4) _____

(5) _____   (6) _____

(7) _____   (8) _____

(9) _____   (10) _____

Other _____

Emergency decontamination will include the following stations: _____

_____

_____

_____

The following decontamination equipment is required: _____

_____

_____

_____

(detergent and water) will be used as the decontamination solution.

J. Site Safety and Health Plan

1. Site Safety Officer

(name) is designated site safety officer and is directly responsible to the project team leader for safety recommendations on site.

2. Emergency Medical Care

(names of qualified personnel) are the qualified EMTs on site. (medical facility name) at (address), phone _____ is located _____ minutes from this location. (name of person) was contacted at (time) and briefed on the situation, the potential hazards, and the substances involved. A map of alternative routes to this facility is available at (normally command post).

Local ambulance service is available from (name of service) at phone number _____. Their response time is _____ minutes. Whenever possible, arrangements should be made for on-site standby.

First aid equipment is available on-site at the following locations:

    First aid kit   _____

    Emergency eye wash   _____

    Emergency shower   _____

    Other _____   _____

Emergency medical information for substances present:

Substance	*Exposure Symptoms*	*First Aid Instructions*
_____	_____	_____
_____	_____	_____
_____	_____	_____
_____	_____	_____
_____	_____	_____

List of emergency telephone numbers:

Agency/Facility	*Phone #*	*Contact*
Police _____		_____
Fire _____		_____
Hospital _____		_____
Airport _____		_____
Public Health Advisor _____		_____

3. Environmental Monitoring

The following environmental monitoring instruments shall be used on site (if required) at the specified intervals.

    Combustible Gas Indicator   - continuous/hourly/daily/other _____

    Oxygen (O_2) Monitor   - continuous/hourly/daily/other _____

Colorimetric Tubes	- continuous/hourly/daily/other _____
_____	_____
_____	_____
HNu (PID)/OVA (FID)	- continuous/hourly/daily/other _____
Toxic Gas/Vapor Meters	- continuous/hourly/daily/other _____
_____	_____
Other _____	- continuous/hourly/daily/other _____
_____	_____

4. **Emergency Procedures** (modified as required for incident)

The following standard emergency procedures will be used by on-site personnel. the site safety officer shall be notified of any on-site emergencies and be responsible for ensuring that the appropriate procedures are followed.

Personnel injury in the exclusion zone: upon notification of any injury in the exclusion zone, the designated emergency signal _____ shall be sounded. All site personnel shall assemble at the designated gathering point. The rescue team will enter the exclusion zone (if required) to remove the injured person to the hotline. The site safety officer and project team leader should evaluate the nature of the injury, and the affected person should be decontaminated to the extent possible prior to movement to support zone. The on-site EMT shall initiate the appropriate first aid, and contact should be made for an ambulance and with the designated facility (if required). No person shall reenter the exclusion zone until the cause and nature of the injury are determined.

Personnel injury in the support zone: Upon notification of an injury in the support zone, the project team leader and site safety officer will assess the nature of the injury. If the cause of the injury or the loss of the injured person does not affect the performance of site personnel, operations may continue, with the on-site EMT initiating the appropriate first aid and necessary follow-up as stated above. If the injury increases the risk to others, the designated emergency signal _____ shall be sounded and all site personnel shall move to the designated gathering point for further instructions. Activities on site will stop until the added risk is removed or minimized.

Fire/explosion: Upon notification of a fire or explosion on site, the designated emergency signal _____ shall be sounded and all site personnel shall assemble at the designated gathering point. The fire department shall be alerted and all personnel moved to a safe distance from the involved area.

Personal protective equipment failure: If any site worker experiences a failure or alteration of protective equipment that affects the protection factor, that person and his/her buddy shall immediately leave the exclusion zone. Reentry shall not be permitted until the equipment has been repaired or replaced.

Other equipment failure: If any other equipment on site fails to operate properly, the project team leader and site safety officer shall be notified and then determine the effect of this failure on continuing operations on site. If the failure affects the safety of personnel or prevents the completion of the work plan tasks, all personnel shall leave the exclusion zone until the situation is evaluated and appropriate actions taken.

Emergency escape routes: The following escape routes are designated for use in those situations where egress from the exclusion zone cannot occur through the decontamination line:

_____(describe alternate routes to leave area in emergencies)_____

_____

_____

In all situations, when an on-site emergency results in evacuation of the exclusion zone, personnel shall not reenter until:

1. The conditions resulting in the emergency have been corrected.
2. The hazards have been reassessed.
3. The site safety plan has been reviewed or revised.
4. Site personnel have been briefed on any changes in the site safety plan.

5. Personal Monitoring

The following personal monitoring will be in effect on-site:

Personal exposure sampling: (describe any personal sampling programs being carried out on site personnel. This would include use of sampling pumps, air monitors, dosimeters, etc.) _____

Medical monitoring: The expected average temperature (°F) will range between _____ If it is determined that heat stress monitoring is required (mandatory over 70°F) the following procedures shall be followed: (describe procedures in effect, i.e., monitoring body temperature, body weight (water loss), pulse rate, etc.) _____

_____

_____

_____

All site personnel have read the above plan and are familiar with its provisions.

Site Safety Officer	_____(name)_____	_____(signature)_____
Project Team Leader	_____	_____
Other Site Personnel	_____	_____
	_____	_____
	_____	_____
	_____	_____

REVIEW QUESTIONS

1. List the emergency equipment that should be available to handle emergencies.
2. When should the site safety plan be updated or modified?
3. What air monitoring requirements should be included?

REFERENCES

EPA, *Standard Operating Safety Guides*, U.S. Government Printing Office, Washington, D.C., 1984.

NIOSH/OSHA/USCG/EPA, *Safety and Health Guidance Manual for Hazardous Waste Site Activities*, U.S. Government Printing Office, Washington, D.C., 1985.

OSHA 29 CFR 1910.120, *Hazardous Waste Operations and Emergency Response (SARA)*, U.S. Government Printing Office, Washington, D.C., 1989.

CHAPTER 12

MEDICAL SURVEILLANCE PROGRAM

Medical surveillance is the preplacement and routine examination of employees at risk for material impairment from exposure to hazardous substances or harmful physical agents. Such examination will help to detect potential work-related signs and symptoms of such exposure.

If such signs or symptoms are detected, the employee and his/her employer are notified so that corrective action may be promptly implemented to avoid further risk.

Another purpose is to identify any limitations of affected employees with regard to the wearing of respirators and other personal protective equipment.

CHAPTER OBJECTIVES

When you have completed this chapter, you will know

- The types of exams and how often they are given
- What is included in the program
- Who is covered and who pays for the exam
- How notification is handled

- What records must be kept and for how long
- How accession will be handled

12.1 THE MEDICAL EXAMINATION

12.1.1 Types of Exams and When They Are Required

Preplacement Exams These must be performed to detect preexisting conditions and to obtain a baseline. The examination must take place *prior* to the employee's *initial* assignment following date of hire.

Annual Exams These must be provided to employees every year following the year of preplacement exam and in which the following conditions are met: 30 days at or above an established permissible limit (PEL), or 30 or more days per year wearing a respirator. More frequent examinations may be scheduled if the examining physician feels that the increased frequency is warranted

Termination-of-Employment Exams These exams occur whenever

- An employee terminates employment with an employer
- Reassignment to a different work area where the employee would not be covered, if he/she has not had an examination within the last 6 months.

Following an Incident Involving Exposure Signs or Symptoms An examination must occur as soon as possible upon notification to the employer that the employee has developed signs or symptoms indicating possible overexposure to hazardous substances or health hazards, or that the employee has been exposed at or above the established exposure levels in an emergency situation.

Emergency These exams are essential for employees injured due to overexposure from an emergency incident involving hazardous substances or health hazards, at concentrations above the established permissible exposure limits, without use of the necessary personal protective equipment.

All medical exams and procedures required in this section shall be performed by or under the supervision of a licensed physician. The ex-

amining physician may require more frequent examinations than indicated above, if the specific situation warrants.

12.1.2 The Examination Content

Medical examinations must include

- A medical history
- A work history

The medical and work history may be an updated history, if one is already in the employee's medical file.

The medical and work history should have a special emphasis on symptoms relating to the handling of hazardous substances and the fitness for duty. The latter includes the ability to wear respiratory and personal protective equipment under temperature extremes likely to be experienced on-site.

The content of exams or consultations made available to employees under this section shall be determined by the examining physician.

12.1.3 Coverage and Payment for Medical Examinations

Those legally required by the standard to be part of a medical surveillance program are

- Members of Hazmat teams CFR.120(1)(4)(ii)
- Employees engaged in site cleanup and remediation activities, CFR 1910.120(b)
- Other employees covered by CFR1910.120(0)(3)

All employees who engage in any aspect of hazardous waste handling *should* be a part of a medical surveillance program.

The required exams and procedures must be provided by the employer

- Without cost to the employee
- Without loss of pay to the employee
- At a reasonable place and time

12.1.4 Notification of Results

The Written Physician's Opinion

1. The employer must obtain and furnish the affected employee with a copy of the *written opinion* of the examining physician.

2. The written opinion of the examining physician must include at least the following:
   a. The results of the medical examination and test *if requested by the employee*
   b. The physician's opinion as to whether the employee has any detected medical conditions that would place the employee at increased risk of material health impairment, arising from work in hazardous waste operations or emergency response or from respirator use as required by 1910.134
   c. The physician's recommended limitations upon the employee's assigned work
   d. A statement that the employee has been informed by the physician of the results of the medical examination and any medical conditions that require further examination or treatment
3. The written opinion obtained by the employer shall not reveal specific findings or diagnoses unrelated to occupational exposure.

Record Keeping

1. The *employer* must keep an accurate record of medical surveillance required by this standard (29 CFR 1910.120).
2. Records of medical surveillance must be preserved and retained by the *employer* for 30 years plus the duration of employment of each affected employee [1910.20(d)(1)(i)].
3. The employer must retain
   a. The name and social security number of the employee
   b. The physician's written opinions, recommended limitations, and results of examinations and tests
   c. Any employee medical complaints related to exposure to hazardous substances
   d. A copy of the information provided to the examining physician by the employer, with the exception of the standard and its appendices

12.1.5 Employee Access to Medical Records

OSHA Standard 1910.120 requires that employers maintain and preserve medical records, including those required under the hazardous waste standard. This standard also applies to *employee exposure* records. These must be maintained for 30 years by the employer.

An employee need only request from his or her employer access to their medical and/or exposure records. The employer must then make this information available within 15 working days.

The employer cannot charge any fee for the first set of copies of the requested records. Methods by which access can be given by the employer are as follows:

- Provide a copy of the records at no cost to the employee
- Provide the necessary copying facilities at no cost to the employee
- Loan the record to the employee to allow them to copy it

What is meant by employee *exposure records*? Employee exposure records include monitoring data for toxic substances or harmful physical agents such as

- Records of the employee's past or present exposure
- Exposure records of other employees with past or present job assignments similar to those of the employee
- Records pertaining to workplaces or working conditions to which the employee is being assigned or transferred

Exposure records include

- Environmental (workplace) monitoring: personal, environmental, grab, wipe, or other sampling techniques, including the related collection and analytical methods
- Biological monitoring results
- Material safety data sheets

12.1.6 Employee Rights to Examination Records

Whenever an employee requests his/her medical records, a physician representing the employer may recommend

- Consultation with the physician to review and discuss the records requested
- That the requesting employee accept a summary in lieu of the medical records requested
- Release of the requested records only to a physician or other designated representative

12.2 TESTS AND PROTOCOLS

The tests listed in Tables 12.1 to 12.3 are performed when required on site workers during their preplacement exam and during other required examinations.

TABLE 12.1 ■ Blood Chemistry Profile

This profile includes 25 tests.

Glucose: A measure of sugar levels in the blood. All organs of the body require glucose to function, and they require normal glucose levels to function best. Abnormal glucose levels may reflect abnormalities in the liver or pancreas. High glucose levels may be seen in diabetes.

Sodium/Chloride Potassium: Components of mineral salts. The levels in the body are controlled by the kidneys and adrenal glands. These mineral salts are important for the proper functioning of nerves, muscles, the heart, and other organs. Levels may be reduced by diuretic medications.

BUN (Blood Urea Nitrogen): A waste product that is excreted by the kidneys. High values may reflect abnormal kidney function. The BUN may also be affected by alterations in dietary protein or fluid intake.

Creatinine: Another waste product that is excreted by the kidney. Unlike BUN, the creatinine level is not significantly affected by alterations in protein intake.

BUN/Creatinine Ratio: In the event that an abnormal BUN or creatinine level is found, the BUN/creatinine ratio may provide information that could help to identify the cause of the abnormality.

Uric Acid: A metabolite that is normally excreted in the urine. High values may be associated with kidney problems and gout. Diuretic medications and, sometimes, lead toxicity may increase uric acid levels.

Calcium and Phosphate: There are minerals that are necessary for normal cellular functions and important components of bones and teeth. The levels of calcium and phosphate in the body are controlled by the parathyroid glands and the kidneys and may be influenced by dietary factors.

Magnesium: Another important mineral. Magnesium plays a key role in proper nerve and muscle function.

Cholesterol and Triglycerides: These are fatty substances that are present in the blood. Elevated cholesterol levels are associated with an increased risk of heart disease. Reduction of elevated cholesterol levels may reduce heart disease risk.

TABLE 12.1 ■ Continued

Total Protein, Albumin, Globulin, and the Albumin/Globulin Ratio: The proteins present in the blood include albumin and the globulins, and these taken together make up the total protein. The proteins play an important role in various metabolic processes and in the transport of materials throughout the body. Measurements of these proteins provide a valuable index of general health, nutrition status, and the function of various organs.

Bilirubin (Total and Direct): Bilirubin is a by-product of the metabolism of red blood cells. It is excreted by the liver and acts as the primary pigment in bile. Abnormal values may reflect abnormal liver function or accelerated destruction of red blood cells.

Alkaline Phosphatase: An enzyme found in the liver and bones. Elevated levels may reflect disease or injury of these organs.

GGT (G-Glutamyl Transpeptidase): An enzyme found in the liver. Elevated levels may reflect liver dysfunction or damage.

SGO and SGP Transaminases: These are enzymes that participate in metabolic processes in various organs, including the liver, muscles, and the heart. Elevated levels may indicate damage to cells that have allowed the release of an excessive amount of these enzymes into the blood.

LDH (Lactate Dehydrogenase): An enzyme that is present in all of the cells of the body. Elevated levels may occur in the event of damage to cells.

Iron: An important mineral that is used in various ways in the body. It is particularly important in the production of hemoglobin and red blood cells. Low levels may be associated with anemia. High levels may be associated with other diseases.

12.3 EXAMPLES OF HAZARDOUS CHEMICALS REQUIRING MEDICAL SURVEILLANCE

12.3.1 Benzene

Uses Benzene is an intermediate in the production of styrene, phenol, cyclohexane, and other organic chemicals. It is also used in the manufacture of detergents, pesticides, solvents, explosives, pharmaceuticals, dyestuffs and paint removers, and gasoline.

TABLE 12.2 ■ Hematology Profile— The "CBC" (Complete Blood Count)

This profile includes 11 tests.

WBCs (White Blood Cells): This is a count of the number of white blood cells present in the blood. White blood cells are important for fighting infection. Abnormal counts may occur with infection, toxic exposures, and various other diseases. The different types of WBCs are counted in a test called the *differential count.*

RBCs (Red Blood Cells): This is a count of the number of red blood cells present in the blood. Red blood cells carry oxygen from the lungs to all the tissues of the body. Low counts may occur in anemia. High counts may be associated with a number of diseases.

HGB (Hemoglobin): Hemoglobin is the substance in the red blood cell that actually "carries" the oxygen. The hemoglobin level may be low in anemia.

HCT (Hematocrit): This is similar to the red blood cell count except that it measures the amount of red blood cells present as a percentage of the total blood volume.

MCV (Mean Cell Volume): This is a measure of the average size of the red blood cells. Many diseases including vitamin and iron deficiencies can alter the size of the red blood cells.

MCH (Mean Cell Hemoglobin) and MCHC (Mean Cell Hemoglobin Concentration): These measure the average amount of hemoglobin and the concentration of hemoglobin in the red blood cells.

RDW (Red Blood Cell Width): This is a measure of the range of different sizes of red blood cells.

Differential Count: This counts the numbers of each of the different types of white blood cells present in the blood. The different types of white blood cells include the neutrophils, lymphocytes, monocytes, eosinophils, and basophils. Each type of white blood cell serves a specific set of functions, and different ones may show abnormal counts under different conditions.

Platelet Count: The platelet is a blood component that plays a critical role in the body's blood clotting system. The number of platelets may be altered in various disease states.

MPV (Mean Platelet Volume): This is a measure of the average size of the platelets.

EXAMPLES OF HAZARDOUS CHEMICALS ■ 377

TABLE 12.3 ■ Examination Protocols

Component	Preplacement	Periodic	Exit
Histories and Examinations			
Hazardous waste history questionnaire (comprehensive baseline type)	X		
Hazardous waste history questionnaire (modified annual type)		X	X
Physical with neurological	X	X	X
Testing Procedures			
Chest x-ray	X	X[a]	X
Electrocardiogram	X	X	X
Pulmonary function testing	X	X	X
Vision screening	X	X	X
Blood chemistry profile	X	X	X
Hematology profile (CBC)	X	X	X
Urinalysis	X	X	X
RBC cholinesterase	X	X	X
PCB level	X	X	X
Blood lead level	X	X	X
Urine porphyrins	X	X	X
Urine heavy metals	X	X[b]	X

[a]Every 3 years or as indicated by physician.
[b]The need for this test may depend on exposure history.

Metabolism Absorption of benzene occurs by

- Inhalation of vapor (primary route)
- Ingestion
- Being poorly absorbed through intact skin

Excretion of benzene occurs by

- Conversion in liver to phenols and elimination in urine
- Being exhaled in breath

Metabolic Effects

Acute
- Narcosis, acute intoxication with the following symptoms:

    Dizziness

    Giddiness

    Euphoria

    Convulsions

    Stupor

    Coma

    Death

- Mucous membrane irritation

Chronic
- Bone marrow depression from phenols producing anemia, pancytopenia, and aplastic anemia
- Myelomonocytic leukemia
- Liver disease
- Chronic dermatitis from defatting (cumulative irritant contact dermatitis)

Treatment
- Removal
- Avoid epinephrine, alcohol, digestible fat and oils
- Symptomatic and supportive

Engineering Controls All systems and operations using benzene should be completely enclosed. Exposures should be reduced to as low as possibly achievable.

Medical Surveillance Evaluations every 6 months should include CBC's reticulocyte count and platelet count, liver function studies, urinalysis, and urine total phenol.

12.3.2 Toluene and Xylene

Uses Toluene and xylene are used as solvents and organic intermediates.

Metabolism Absorption of these chemicals occurs by

- Inhalation of vapors (primary mode of entry)
- Ingestion
- Being poorly absorbed through intact skin

Excretion of these chemicals occurs by

- Being exhaled in breath
- Toluene excreted in urine as hippuric acid
- Xylene excretion believed to be similar to toluene

Metabolic Effects

Acute

- Narcosis and/or acute intoxication with the following symptoms:

  Dizziness

  Giddiness

  Euphoria

  Convulsions

  Stupor

  Coma

  Death
- Mucous membrane irritation—xylene
- Reversible liver and kidney impairment—xylene

Chronic

- Reversible liver and kidney impairment—xylene
- Habituation—toluene glue sniffing
- Dermatitis—both
- Mild anemia, leukopenia—possibly from benzene contamination

Treatment

- Removal from exposure
- Avoid epinephrine, alcohol, digestible fat and oils

- Symptomatic and supportive

Medical Surveillance Medical surveillance includes preplacement and periodic exams with emphasis on skin, central nervous system (CNS), liver, and kidney function studies and a CBC if there is a chance of benzene contamination of either compound.

12.3.3 Formaldehyde (HCHO)

Uses Formaldehyde is used as an ionizing solvent and in the manufacture of formic esters, resins, plastics, leather, rubber, metals, and wood. It also is used for disinfection in dialysis units.

Metabolism Absorption of formaldehyde occurs by

- Inhalation (primary route)
- Skin absorption

Excretion of formaldehyde occurs by

- Excretion as formic acid in urine, formaldehyde in blood

Metabolic Effects

Acute
- Irritating to eyes, mucous membranes, and upper respiratory tract

Chronic
- Chronic dermatitis from defatting and sensitization can occur
- Possible cause for occupational asthma (late asthmatic reactions have occurred)
- Suspected human carcinogen—has been associated with increased cancer deaths
- Brain damage
- Leukemias

Treatment
- Removal from contaminated environment
- Reduce exposures by use of engineering controls

Medical Surveillance Medical surveillance includes preplacement and periodic exams with special attention to eyes, mucous membranes, respiratory system, and skin.

12.3.4 Ketones

Acetone	CH_3COCH_3
Diacetone	$(CH_3)_2COHCH_2COCH_3$
Methyl ethyl ketone (MEK)	$CH_3COCH_2CH_3$
Methyl *n*-propyl ketone	$CH_3CH_2CH_2COCH_3$
Methyl *n*-butyl ketone (MBK)	$CH_3COCH_2CH_2CH_2CH_3$
Methyl isobutyl ketone	$(CH_3)_2CHCH_2COCH_3$

Uses Ketones are used as low-cost solvents for resins, lacquers, oils, fats, collidion, cotton, cellulose acetate, and others. They also are used as chemical intermediates in production of explosives, paint, plastics, celluloid, cement, rubber, and lubricating oils. In addition, several ketones are used as hydraulic fluids, metal cleaning compounds, quick drying inks, airplane dopes, paint remover and dewaxers.

Metabolism Absorption of ketones occurs by

- Inhalation (primary route)
- Skin absorption

Excretion of ketones occurs by

- Being exhaled in breath
- Excretion in urine

Metabolic Effects

Acute
- Narcosis and/or acute intoxication with the following symptoms:

  Dizziness

  Giddiness

  Euphoria

  Convulsions

  Stupor

  Coma

  Death
- Respiratory and mucous membrane irritation

Chronic
- Peripheral neuropathy (MEK & MBK)
- Chronic dermatitis from defatting of the skin

Treatment
- Removal from contaminated environment
- Symptomatic and supportive care

Medical Surveillance Medical surveillance includes preplacement and periodic exams with emphasis on skin and nervous system. Liver evaluation may be necessary because this group is often mixed with other hepatotoxic substances

12.3.5 Chlorinated Aliphatic Hydrocarbons

Carbon tetrachloride	CCl_4
Chloroform	$CHCl_3$
Methyl chloroform	Cl_3CCH_3
Trichloroethylene (TCE)	$CCl_2=CHCl$
Perchloroethylene (Perc)	$CCl_2=CCl_2$

Uses All are used as solvents and chemical intermediates.

Metabolism
Absorption

	Lungs	GI Tract	Skin
Carbon tetrachloride	+++	++	+
Chloroform	+++	++	+
Methyl chloroform	+++	++	+
Trichloroethylene	+++	+	+
Perchloroethylene	+++	+	+

Excretion

	Lungs	Urine
Carbon tetrachloride	+++	-
Chloroform	++	+

Methyl chloroform	++	+
Trichloroethylene	+	++
Perchloroethylene	++	+

Metabolic Effects

Acute

- CNS depression—all chlorinated aliphatic hydrocarbons
- Liver damage—all of the above chlorinated aliphatic hydrocarbons cause acute yellow atrophy
- Renal damage—all (Perchloroethylene affects animals only)
- Cardiac sensitization—caused by carbon tetrachloride, methyl chloroform, and TCE
- Mucous membrane irritation—all chlorinated aliphatic hydrocarbons cause irritation at high vapor concentrations or direct contact

Chronic

- Dermatitis (defatting of the skin)—all chlorinated aliphatic hydrocarbons
- Liver damage (chloroform and perchloroethylene in animals)
- Renal damage (chloroform and perchloroethylene in animals)
- Flushing after drinking alcohol and TCE
- Various CNS effects—carbon tetrachloride and TCE
- Peripheral neuropathy—perchloroethylene and TCE
- Habituation—TCE
- Animal carcinogenesis

    Carbon tetrachloride (liver)

    Chloroform (liver and kidney)

    Methyl chloroform (?)

    TCE and perchloroethylene (liver)
- Animal teratogenesis—carbon tetrachloride, chloroform
- Animal fetotoxic effects—perchloroethylene

Treatment

- Remove from exposure
- Avoid alcohol and epinephrine
- Symptomatic and supportive care

TABLE 12.4 ■ **Department of Labor, Occupational Safety, and Health Administration Regional Offices**

Region I
(CT[a], MA, ME, NH, RI, VT[a])
133 Portland Street
1st Floor
Boston, MA 02114
Telephone: (617) 565-7614

Region II
(NJ, NY[a], PR, VI[a])
201 Varick Street
Room 670
New York, NY 10014
Telephone: (212) 337-2378

Region III
(DC, DE, MD[a], PA, VA[a], WV)
Gateway Building, Suite 2100
3535 Market Street
Philadelphia, PA 19104
Telephone: (215) 596-1201

Region IV
(AL, FL, GA, KY[a], MS, NC[a], SC[a], TN[a])
1375 Peachtree Street, N.E.
Suite 587
Atlanta, GA 30367
Telephone: (404) 347-3573

Region V
(IL, IN[a], MI[a], MN[a], OH, WI)
230 South Dearborn Street
Room 3244
Chicago IL 60604
Telephone: (312) 353-2220

Region VI
(AR, LA, NM[a], OK, TX)
525 Griffin Street
Room 602
Dallas, TX 75202
Telephone: (214) 767-4731

Region VII
(IA[a], KS, MO, NE)
911 Walnut Street
Room 406
Kansas City, MO 64106
Telephone: (816) 426-5861

Region VIII
(CO, MT, ND, SD, UT[a], WY[a])
Federal Building, Room 1576
1961 Stout Street
Denver, CO 80294
Telephone: (303) 844-3061

Region IX
(AZ[a], CA[a], HI[a], NV[a])
71 Stevenson Street
Room 415
San Francisco, CA 94105
Telephone: (415) 995-5672

Region X
(AK[a], ID, OR[a], WA[a])
Federal Office Building
909 First Avenue
Room 6003
Seattle, WA 98174
Telephone: (206) 442-5930

[a]These states and territories operate their own OSHA-approved job safety and health programs (the Connecticut and New York plans cover public employees only, and OSHA currently is exercising concurrent private-sector federal enforcement authority in California).

- Hypothermia (after acute CNS, symptoms subside to prevent hepatic and renal damage—experimental)

Medical Surveillance Medical surveillance includes preplacement and periodic exams with special attention to skin, liver, kidney, CNS, and cardiovascular system (liver and renal functions tests may be required).

If the employee, employer or physician need more information or guidance in establishing or improving a medical surveilance program, they can contact the regional OSHA offices which are listed in Table 12.4.

REVIEW QUESTIONS

1. How long must medical records be kept after employee termination?
2. What is meant by exposure records? What do they include?

REFERENCES

NIOSH/OSHA/USCG/EPA, *Safety and Health Guidance Manual for Hazardous Waste Site Activities*, U.S. Government Printing Office, Washington, D.C., 1985.

OSHA 29 CFR 1910.120, *Hazardous Waste Operations and Emergency Response (SARA)*, Washington, D.C., U.S. Government Printing Office, 1989.

Waxman, Michael F. ed., *Hazardous Waste Site Operations—40 Hour OSHA Training Workshop*, University of Wisconsin, August 29–September 1, 1994.

CHAPTER 13

RISK ASSESSMENT IN SUPERFUND SITE REMEDIATION

Risk assessment is the cornerstone of environmental decision-making. Despite this role as the scientific foundation for most EPA regulatory actions, risk assessment means different things to different people. Some points of controversy involve the interpretation of scientific studies. Others have to do with science policy issues. Still others center on distinctions between risk assessment and risk management.

The scope and nature of risk assessments range widely—from broadly based scientific conclusions about air pollutants such as lead or arsenic that affect the nation as a whole, to site-specific findings concerning these same chemicals in a local water supply. Some assessments are retrospective, focusing on injury after the fact—for example, the kind and extent of risks at a particular Superfund site. Others seek to predict possible future harm to human health or the environment—for example, the risks expected if a newly developed pesticide is approved for use on food crops.

In short, risk assessment takes many different forms, depending on its intended scope and purpose, the available data and resources, and other factors. It involves many different disciplines and specialists with different kinds and levels of expertise, representing many different organizations. Moreover, risk assessment approaches differ somewhat in line with differences in environmental laws and related regulatory programs.

13.1 RISK ASSESSMENT AND RISK MANAGEMENT

Risk assessment and risk management are closely related but different processes, with the nature of risk management decisions often influencing the scope and depth of a risk assessment.

In simple terms, risk assessment asks, "How risky is the situation?" and risk management then asks, "What shall we do about it?"

We use the term "risk assessment" to mean the process by which scientific data are analyzed to describe the form, dimension, and characteristics of risk—that is, the likelihood of harm to humans or to the environment. Risk management, on the other hand, is the process by which the risk assessment is used with other information to make important decisions such as cleanup goals and regulatory approaches.

13.2 DISCIPLINES INVOLVED IN THE PROCESS

Environmental risk assessment is a multidisciplinary process. It draws on data, information, and principles from many scientific disciplines including biology, chemistry, physics, medicine, geology, epidemiology, and statistics, among others. The feature distinguishing risk assessment from the underlying sciences is this: After evaluating individual studies for conformity with standard practices within the discipline, the most relevant information from each of these areas is examined as a whole to describe the risk. This means that individual studies, or even collections of studies from a single discipline, are used to develop risk assessments, but they are not in themselves generally regarded as risk assessments, nor can they alone generate risk assessments.

One way to highlight differences between risk assessment and risk management is by looking at differences in the information content of the two processes. What kind of information, then, is used for risk management but not for risk assessment? In general practice, data on technological feasibility, costs, and economic and social consequences of possible regulatory decisions are critically important for risk management but not for risk assessment. To the extent called for in various statutes, risk managers consider this information together with the outcome of the risk assessment when evaluating risk management options and making environmental decisions.

The risk assessment model put forward by the National Academy of Sciences (NAS) in a 1983 publication called *Risk Assessment in the Federal*

Government: Managing the Process defines four "fields of analysis" that describe the use and flow of scientific information in the risk assessment process.

The following discussion reviews these four fields of analysis. Note at the outset that each phase employs different parts of the information base. For example, *data evaluation* relies primarily on data from the biological and medical sciences. *Toxicity assessment* then uses these data in combination with statistical and mathematical modeling techniques, so that the second phase of the risk analysis builds on the first.

13.3 THE FOUR FIELDS OF ANALYSIS
13.3.1 Data Evaluation

The objective of this phase is to determine whether the available scientific data describe a causal relationship between an environmental agent and demonstrated injury to human health or the environment. In humans, the observed injury may include such effects as birth defects, neurologic effects (nerve damage), or cancer. Ecological hazards might result in fish kills, habitat destruction, or other effects on the natural environment.

Information on the agent responsible for the effects may come from laboratory studies in which test animals were deliberately exposed to toxic materials, or from other sources such as chemical measurements in the workplace. In addition, studies on a pollutant's effects on genetic material or metabolism, along with comparison of such effects in humans and experimental test systems, may be part of the analysis.

The principal question is whether data from populations in which effects and exposure are known to occur together suggest a potential hazard for other populations under expected conditions of exposure to the agent under study. If a potential hazard is identified, three other analyses become important for the overall risk assessment, as discussed below.

13.3.2 Toxicity Assessment

The dose–response analysis is designed to establish the quantitative relationship between exposure (or dose) and response in existing studies in which adverse health or environmental effects have been observed. The dose–response analysis is based mainly on two extrapolations. One extrapolation uses the relatively high exposure levels in most laboratory studies (or, for example, human studies at relatively high workplace levels) to estimate the probable magnitude of the effect in the same population at lower environmental levels where little or no data are available.

The other extrapolation entails looking for the expected level of response in humans, or in animals or plants in nature, based on comparisons of data from laboratory and natural test systems. As explained later, each extrapolation involves numerous scientific uncertainties and assumptions, which in turn involve policy choices.

The number produced in the toxicity assessment—perhaps a cancer risk value or a reference—is sometimes regarded as a risk assessment because it describes important information from animal and human studies. However, risk assessment is complete only when human exposure assessment information is joined with dose–response analysis and all relevant information to characterize the risk.

13.3.3 Exposure Assessment

The exposure analysis (i.e., part of the exposure assessment) moves the assessment from the study of known populations (laboratory or epidemiologic) in which dose (exposure) and response occur together, to the task of identifying and characterizing exposure in other potentially exposed populations. These populations may be as general as the nation as a whole for certain widely distributed materials (e.g., contaminated food), or as limited as certain occupations or user groups (e.g., pesticide applicators or site workers). Questions raised in the exposure analysis concern the likely sources of the pollutant (e.g., leaking underground storage tanks, incinerator discharge, factory effluent, pesticide application), its concentration at the source, its pathways (air, water, food) from the source to target populations, and actual levels impacting target organisms.

The exposure analysis relies on many very different kinds of information, some based on actual measurements and some developed using mathematical models. Measurements of the kind and quantity of a pollutant in various environmental media and, when available, in human, plant, and animal tissues are used to project expected exposure levels in individuals, populations, or both. The exposure analysis also develops "lifestyle" data to identify and describe populations likely to contact a pollutant. For example, if a chemical that causes birth defects in test animals contaminates apples, the exposure analysis would consider such "lifestyle" information as the number of women of childbearing age who eat apples, how often they eat this food, and in what quantities. To complete the exposure analysis, the lifestyle information is combined with information on how much chemical, probably measured at very low levels, remains in apples when sold for consumption.

If the estimated exposure for an environmentally exposed population is significantly smaller than the lowest dose producing a response in the study population, then, the likelihood of injury to exposed humans is smaller. However, if the estimated exposure is significantly greater than the lowest dose, then the likelihood of injury is greater.

13.3.4 Risk Characterization

Although each of the preceding analyses examines all relevant data and information to describe hazard or toxicity or exposure, none reaches conclusions about the overall *risk*. That task is reserved for the final analysis, where important information, data, and conclusions from each of the preceding analyses are examined together to characterize risk—that is, to fully describe the expected risk by examining the exposure predictions for real-world conditions in light of the toxicity assessment information from animals, people, and special test systems.

Risk characterization—the product of the risk assessment—is much more than a number or value. While the risk is often stated as a bare number—for example, "a risk of 10^{-6}" or "one in a million new cancer cases"—the analysis involves substantially more information, thought, and judgment than the numbers express.

What the Numbers Mean Risk values are often stated as a number. When the risk concern is cancer, the risk number represents a probability of occurrence of additional cancer cases. For example, such an estimate for contaminant X might be expressed as 1×10^{-6}, or simply 10^{-6}. This number can also be written as 0.000001, or one in a million—meaning one additional case of cancer projected in a population of one million people exposed to a certain level of contaminant X over their lifetimes. Similarly, 5×10^{-7}, or 0.0000005, or five in 10 million, indicates a potential risk of five additional cancer cases in a population of 10 million people exposed to a certain level of the contaminant. These numbers signify incremental cases above the background cancer incidence in the general population. American Cancer Society statistics indicate that the background cancer incidence in the general population is one in three over a lifetime.

If the effect associated with contaminant X is not cancer but another health effect, perhaps neurotoxicity (nerve damage) or birth defects, then numbers are not typically given as probability of occurrence, but rather as levels of exposure estimated to be without harm. This often takes the form of a reference dose (RfD). An RfD is typically expressed in terms of milligrams (contaminant) per kilogram of body weight per day—for example, 0.004 mg/kg/day. Simply described, an RfD is a rough estimate of daily exposure to the human population (including sensitive subgroups) that is likely to be without appreciable risk of deleterious effects during a lifetime. The uncertainty in RfD may be one or several orders of magnitude (i.e., multiples of 10).

What's in a number? The important point to remember is that the numbers by themselves don't tell the whole story. For instance, even though the numbers are identical, a cancer risk value of 10^{-6} for the "average exposed person" (perhaps someone exposed through the food supply) is not the same thing as a cancer risk of 10^{-6} for a "most exposed individual" (perhaps someone exposed from living or working in a highly contaminated area). It's important to know the difference. Omitting the qualifier "average" or "most exposed" incompletely describes the risk and would mean failure in risk communication.

A numerical estimate or value is only as good as the data on which it is based. Just as important as the *quantitative* aspects of risk characterization (the risk numbers), then, are the *qualitative* aspects. How extensive is the database supporting the risk assessment? Does it include human epidemiological data as well as experimental data? Does the laboratory database include test data on more than one species? If multiple species were tested, did they all respond similarly to the test substance? What are the "data gaps," the missing pieces of the puzzle? What are the scientific uncertainties? What science policy decisions were made to address these uncertainties? What working assumptions underlie the risk assessment? What is the overall confidence level in the risk assessment? All of these qualitative considerations are essential to deciding what reliance to place on a number and to characterizing a potential risk.

13.4 RISK ASSESSMENT IN SUPERFUND

The major decisions at a Superfund site can be made after answering these basic questions:

- Should something be done at this site?
- What should be done?
- When has enough been done?

CERCLA (1980) and SARA (1986) allow the EPA to reduce the risks from hazardous chemical wastes where the environment has been damaged. The National Contingency Plan was written by the EPA to implement and carry out the Superfund law such that it protects the health of both people and the environment.

Different risk assessment approaches are applied in evaluating possible effects of Superfund sites on human health and the environment. In the Superfund program, the application of risk assessment to human health evaluation is well developed and standardized. Neither the EPA nor the Superfund has one standardized environment evaluation methodology, because there are so many different aspects of the environment—individuals, species, ecosystems, natural resources, endangered species, and so on—that may need to be evaluated. Therefore, the focus of this discussion will be the application of risk assessment to human health.

Risk assessment can be a particularly helpful way to view the extent of a problem at a Superfund site. Those managers who work on Superfund sites usually talk about the "baseline risk assessment" conducted during the remedial investigation. In fact, risk assessment is useful to managers throughout the Superfund process, including the

- Pre-remedial program
- Removal program
- Remedial program

Here are some of the questions risk assessment can answer:

- How bad is the site?
- How bad could it become if nothing is done?
- Does the site warrant remedial action?
- How much should be cleaned up for the site to be considered safe?
- What will be the result of the remedial action?

Figures 13.1 and 13.2 show how the risk assessment process relates to the overall Superfund decision-making process.

13.5 BASELINE RISK ASSESSMENT OF THE REMEDIAL INVESTIGATION

The baseline risk assessment of the remedial investigation is the central risk evaluation activity in the Superfund program. As discussed earlier, it is a four-step process (data evaluation, exposure assessment, toxicity assess-

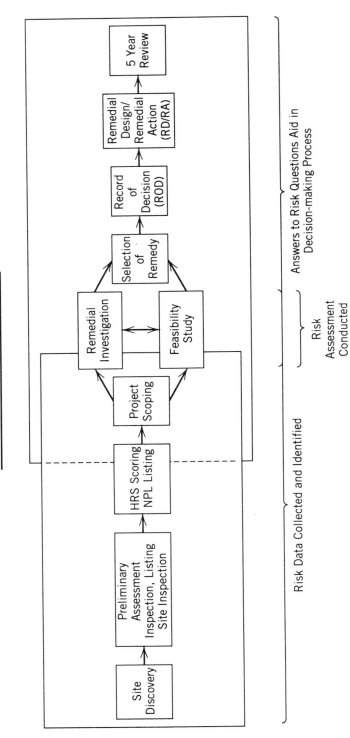

FIGURE 13.1 ■ **Risk assessment activities throughout the Superfund decision-making process.** *Source: Adapted in part from EPA, Risk Assessment Guidance for Superfund: Volume 1—Human Health Evaluation Manual (Part A), 1990.*

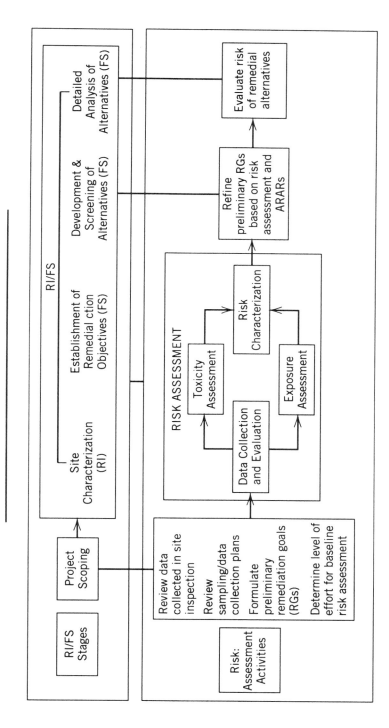

FIGURE 13.2 ■ **A breakdown of the risk assessment activities during the remedial investigation and feasibility study process.** *Source: Adapted in part from EPA, Risk Assessment Guidance for Superfund: Volume 1—Human Health Evaluation Manual (Part A), 1990.*

ment, risk characterization). These four steps are the framework used in each risk evaluation step in the Superfund remedial program.

13.5.1 Data Evaluation

To initiate the baseline risk assessment of the remedial investigation, information on site history and data gathered during the pre-remedial program (preliminary assessment, site inspection, listing site inspection) or a recent site visit must be assembled and used to guide the development of a conceptual understanding of the site. The project management leaders can then prepare a sufficiently detailed workplan for conducting the necessary investigative and analytical tasks. This workplan development occurs in the "scoping" of the remedial investigation.

A critical part of the scoping, in which the risk assessor or toxicologist can assist greatly, is identifying contaminants of significant toxicity and all exposure pathways of concern. In addition to the media paths—soil, air, and drinking water, other paths such as the ingestion of contaminated food or recreation may be very important.

Knowing the toxicity and exposures to be evaluated leads directly to determining a sampling strategy and selecting appropriate analytical methods and related data quality objectives (DQOs) for sampling and analysis. The DQOs specify the quality and quantity of data required to support decisions during the remedial response activities.

The evaluation of data from lab or field analyses identifies chemicals of potential concern at the site.

13.5.2 Exposure Assessment

The exposure assessment looks for complete pathways of exposure to particular types of people on or near the site. The activities of people determine their exposure. Figure 13.3 describes the pathways of exposure to populations on or near the site.

The exposure assessment follows eight steps:

- Characterization of physical setting
- Analysis of contaminant releases
- Identification of potentially exposed populations
- Identification of potential exposure pathways
- Estimation of exposure point concentrations
- Estimation of contaminant intake

13.5 BASELINE RISK ASSESSMENT OF THE REMEDIAL INVESTIGATION ■ 397

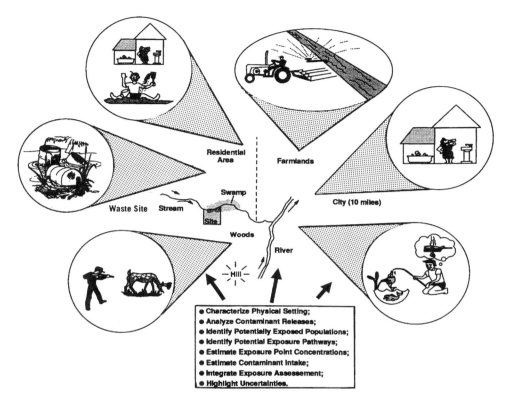

FIGURE 13.3 ■ The exposure assessment looks for complete pathways of exposure of people on or near the site. *Source:* Adapted in part from EPA, *Risk Assessment Guidance for Superfund: Volume 1—Human Health Evaluation Manual (Part A)*, 1990.

- Integration of exposure assessment
- Highlight uncertainties

A site characterization should be undertaken. This includes (a) a detailed site description and (b) a map showing details of on-site and off-site structures, residential neighborhoods, commercial neighbors, street and highway locations, surface water and groundwater flows, and so on.

For specific sites, contaminant releases, potentially exposed populations and the various pathways of exposure are important and should be analyzed in detail:

- Does the site contaminate the groundwater used for drinking and other activities by the commercial and household residents surrounding the site?

- Has dust from the site contaminated playgrounds, parking lots, and residential lots? Will browsing game eat contaminated soils and vegetation?
- Do runoff and groundwater discharges to local streams contaminate the river water or sediments and make their way into fish, crops, or the water supply of cities downstream and then into people?
- Are some populations subjected to combined exposures of several pathways? For example, are local people eating home-grown vegetables and eating fish from contaminated streams?

The baseline risk assessment considers both present exposures and those that might result from current or probable future land use, if no further cleanup action is taken at the site. Figures 13.4 and 13.5 show several illustrations of exposure pathways.

The reasonable maximum exposure (RME) is defined as the highest exposure that is reasonably expected to occur at a site, considering

- Land use
- Intake variables
- Pathway combinations

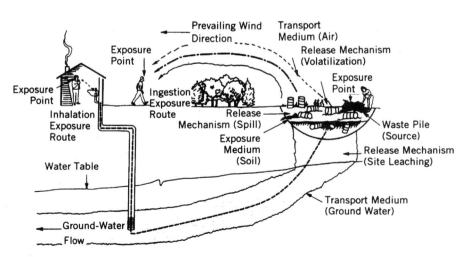

FIGURE 13.4 ■ Illustration of potential exposure pathways prior to remediation. *Source:* Adapted in part from EPA, *Risk Assessment Guidance for Superfund: Volume 1—Human Health Evaluation Manual (Part A)*, 1990.

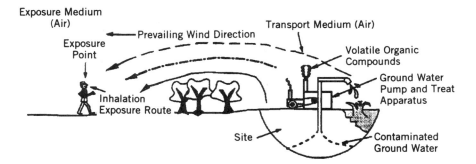

FIGURE 13.5 ▪ **Potential exposure pathways during remedial action.** *Source:* Adapted in part from EPA, *Risk Assessment Guidance for Superfund: Volume 1—Human Health Evaluation Manual (Part A)*, 1990.

RMEs are estimated for individual exposure pathways to generate a "conservative" case (i.e. exposure well above the average case) that is still within the range of possible exposures. Thus people at or near the site will be protected, and cleanup will not be driven by assumed exposures outside the range of possibility.

13.5.3 Toxicity Assessment

As part of the hazard identification step of the toxicity assessment, the risk assessors gather evidence from a variety of sources regarding the potential for a substance to cause adverse health effects (carcinogenic and noncarcinogenic) in humans. These studies include human data (clinical studies and controlled epidemiological investigations), animal data, and supporting data from microorganisms, as well as chemical activity studies on the mode of action of the substance under study.

The toxicologist's or risk assessor's answers to the questions listed in Table 13.1 are required to determine the toxicity of each chemical found at the site.

For chemicals of potential concern, what are the questions that must be answered?

Noncarcinogens A reference dose, or RfD (estimate of the daily exposure for the human population that is likely to be without an appreciable risk of deleterious effects during a portion of a lifetime), is a toxicity value used most often in evaluating noncarcinogenic effects resulting from

TABLE 13.1 ■ Questions That Must Be Answered to Determine Toxicity

Types of effects?	Noncarcinogenic Carcinogenic	Developmental, neurological Acute (single event) Subchronic Chronic (noncarcinogenic)
Qualitative factors?	Animal experiments Human evidence Weight of evidence (A, B, C)[a]	Animal experiments Human evidence
Quantitative factors?	Nonthreshold effects Animal to human extrapolation High to low dose extrapolation	Threshold effects Uncertainty factors Animal to human extrapolation High dose to low dose extrapolation
Toxicity values?	Slope factors Units = $(mg/kg/day)^{-1}$	Reference doses (RfDs) Units = mg/kg/day

[a]Weight-of-evidence classification for chemicals (see discussion under Carcinogens): Group A, human carcinogen; Group B, probable human carcinogen; Group C, possible human carcinogen; Group D, not classifiable; Group E, negative evidence.

exposures at Superfund sites. Various types of RfDs are available depending on the exposure route (oral or inhalation), the critical effect (developmental or other), and the length of exposure being evaluated (chronic, subchronic, or single event).

Threshold Concept For many noncarcinogenic effects, protective mechanisms are believed to exist that must be overcome before an adverse effect is manifested. For example, where a large number of cells perform the same or a similar function, the cell population may have to be significantly depleted before an effect is seen. As a result, a range of exposures exist from zero to some finite value that can be tolerated by the organism without noticeable adverse effects. The toxicity value or RfD is established at the upper bound of the tolerance range. The RfD is generally considered to have uncertainty spanning an order of magnitude or more, and therefore it should not be viewed as a strict demarcation between what level is toxic and nontoxic.

Carcinogens

Nonthreshold Effects For carcinogens, the EPA and the scientific community assume that a small number of molecular events can evoke changes in a single cell that can lead to uncontrolled cellular proliferation and eventually to the clinical manifestations of cancer. Because no level of exposure to a carcinogenic agent is presumed to be risk-free, no threshold exists for these agents.

Toxicologists assess the carcinogenicity (cancer-causing potential) of a substance by studying the effects (number and type of tumors) of different doses (milligrams of substance applied per kilogram of body weight per day), often on laboratory animals.

In a small population of animals, doses that result in a sufficient number of tumors to be statistically significant are often much higher than would occur at the concentration of substance at a Superfund site or somewhere else in the environment. Thus, models are used to extrapolate effects from high (laboratory) dose to low (environmental) dose. Our current understanding of carcinogenicity suggests that even at very low doses, there is some probability of response, that is, there is no threshold concentration above which a response is expected.

In addition to estimating quantitative toxicity values for carcinogens, toxicologists also make qualitative evaluations of the sum total of all studies using a given substance. These evaluations result in a classification system called *weight-of-evidence classifications*. These classifications are used by toxicologists in the risk characterization discussion of the risk assessment.

Weight-of-Evidence Classifications

Group A—Human carcinogen

Group B—Probable human carcinogen

Group C—Possible human carcinogen

Group D—Not classifiable

Group E—Negative evidence

Scientists can adjust the classification up or down, depending on new supporting evidence.

13.5.4 Risk Characterization

Risk characterization is the final step in the risk assessment process. In this step, the toxicity and exposure assessments are summarized and integrated into quantitative and qualitative expressions of risk. To characterize potential noncarcinogenic effects, comparisons are made between projected intakes of substances and toxicity values. To characterize potential carcinogenic effects, probabilities that an individual will develop cancer over a lifetime of exposure are estimated from projected intakes and chemical specific dose–response information. Major assumptions, scientific judgments, and, to the extent possible, estimates of uncertainties are also included in the characterization [for more details see EPA, *Risk Assessment Guidance for Superfund Volume 1—Human Health Evaluation Manual (Parts A and B)*].

Risk characterization serves another function as the bridge between risk assessment and risk management; it is therefore a key step in the ultimate site decision-making process. This step assimilates risk assessment information for the risk manager to be considered alongside other factors important for decision-making such as economics, technical feasibility, and regulatory context.

Most sites being assessed will involve the evaluation of more than one chemical of concern and might include both carcinogenic and noncarcinogenic substances with multiple exposure pathways.

The following steps summarize the risk characterization procedures:

- Combine the toxicity and exposure assessments, calculate the risk for individual carcinogenic and noncarcinogenic effects, and then determine the aggregate risks for multiple substances.
- Determine reasonably anticipated multimedia exposures, combine risks across exposure pathways, and calculate risks for multiple exposure pathways.
- Characterize uncertainties:

Physical setting

Model applicability and assumptions

Multiple substances

Exposure parameter assumptions

Toxicity values

Land use
- Document results including risks and uncertainties.

The Results of Risk Characterization The results should provide a means of placing the numerical estimates of risk and hazard in the context of what is known and what is not known about the site, and in the context of decisions to be made about the selection of remedies. At a minimum, the discussion should include

- Evidence that the key site contaminants were identified with consideration of contaminant concentrations relative to background concentration ranges
- A description of the various types of cancer and other health risks present at the site, distinguishing between known effects in humans and those that are predicted to occur based on animal experiments
- Level of confidence in the quantitative toxicity information used to estimate risks, and presentation of qualitative information on the toxicity of substances not included in the quantitative assessment
- Level of confidence in the exposure estimates for key exposure pathways and related exposure parameter assumptions
- The magnitude of the cancer risks and noncancer risks relative to the Superfund remediation goals (e.g., the cancer range of 10^{-4} to 10^{-7} and noncancer index of 1.0)
- The major factors driving the site risks (e.g., substances, exposure pathways, and pathway combinations)
- The major factors reducing the certainty in the results and the significance of these uncertainties (summation of risk over multiple substances and pathways)
- Exposed population characteristics (e.g., size, age, sex, etc.)

A risk characterization cannot be considered complete unless the numerical expressions of risk are accompanied by explanatory text interpreting and qualifying the results. Risk assessment aids in evaluating remedial alternatives throughout the remedial action process.

13.6 EVALUATING REMEDIAL ALTERNATIVES

Remedial alternatives are evaluated against the nine National Contingency Plan (NCP) criteria listed in Table 13.2.

Applicable or relevant and appropriate requirements (ARARs) are defined as follows:

> "Applicable" requirements are those cleanup standards, standards of control, and other substantive environmental protection requirements, criteria, or limitations promulgated under federal or state law that specifically address a hazardous substance, pollutant, contaminant, remedial action, location, or other circumstance at a CERCLA site. "Relevant and appropriate" requirements are those cleanup standards which, while not "applicable" at a CERCLA site, address problems or situations sufficiently similar to those encountered at the CERCLA site that their use is well-suited to the particular site. ARARs can be action-specific, location-specific, or chemical-specific.

The remedy selected in the Record of Decision (ROD) should best satisfy these criteria. The risk assessment can address several of the above criteria:

TABLE 13.2 ■ Nine Criteria of the National Contingency Plan (NCP)

1. Overall protection of human health and environment 2. Compliance with ARARs (see following discussion)	Threshold criteria
3. Long-term effectiveness and permanence 4. Reduction of toxicity, mobility, or volume 5. Short-term effectiveness 6. Implementability 7. Cost	Primary balancing criteria
8. State acceptance 9. Community acceptance	Modifying Balancing

- The overall protection of human health and the environment is a threshold criterion that the selected remedy must satisfy. Risk assessment for long-term and short-term effectiveness should demonstrate this "protectiveness."
- A risk assessment that relates concentrations of chemicals expected after remediation to residual risks remaining at the site provides a measure of long-term effectiveness.
- A remedy that presents no unacceptable risks during the remedial action meets the short-term effectiveness criterion. Activities at the site may release contaminants, create new pathways of exposure, or even create new chemicals. Risk assessments should evaluate these potential concerns.

Every ROD should identify the contaminants posing risks, target concentrations for cleanup, points of compliance for cleanup in each medium, and the risks that will remain after completion of the remedy if cleanup goals are achieved.

On paper, risk assessment is straightforward, logical, maybe even easy. In practice, each site is unique, and the data will be lacking, uncertainties will require professional judgments, and the best justified and designed remedies will behave in surprising ways. Risk assessment methodology can be applied at any step in the process to help make better, more reasoned decisions.

REVIEW QUESTIONS

1. List the four fields of risk assessment and briefly describe them.
2. **a.** What is the relationship between risk management and risk assessment?
   b. What is risk management?
3. **a.** What are risk values?
   b. How do the values differ for noncarcinogens and carcinogens?
4. Why are exposure assessments performed and what are their functions?
5. What does a risk characterization tell us?

REFERENCES

EPA, *Cleaning Up the Nation's Waste Sites: Markets and Technology Trends*, U.S. Government Printing Office, Washington, D.C., 1993.

EPA, *Risk Assessment Guidance for Superfund: Volume 1—Human Health Evaluation Manual (Part A)*, U.S. Government Printing Office, Washington, D.C., 1989.

EPA, *Risk Assessment Guidance for Superfund: Volume 1—Human Health Evaluation Manual (Part B)*, U.S. Government Printing Office, Washington, D.C., 1991.

EPA, *Risk Assessment Guidance for Superfund: Volume 1—Human Health Evaluation Manual (Part C)*, U.S. Government Printing Office, Washington, D.C., 1991.

EPA, *Risk Assessment Guidance for Superfund: Volume 2—Environmental Evaluation Manual*, U.S. Government Printing Office, Washington, D.C., 1989.

GLOSSARY

Adsorption Penetration of a substance into the body.
ACGIH American Conference of Governmental Industrial Hygienists.
Acid A substance that dissolves or is miscible in water and increases the hydrogen ion concentration. Acids are corrosive. (see also pH).
Acute effect An adverse health effect showing up after a single significant exposure, with severe symptoms developing rapidly and coming quickly to a crisis.
Acute exposure Exposure to a substance in a short time span and generally at high concentrations.
Acute toxicity The adverse (acute) effects resulting from a single dose of, or exposure to, a substance.
Administrative controls Methods of controlling employee exposures by job rotation, work assignments, or time periods away from the hazards.
Adsorption The condensation of gases, liquids, or dissolved substances on the surface of solids.
Aerosols Liquid droplets or solid particles dispersed in air, which are of fine enough particle size (0.01–100 microns) to remain so dispersed for a period of time.
AIHA American Industrial Hygiene Association.
Air monitoring The sampling for and measuring of pollutants in the atmosphere.

Air-purifying respirator Respirators that use filters or sorbents to remove harmful substances from the air.
Air-supplied respirator Respirator that provides a supply of breathable air from a clean source outside the contaminated area.
Aliphatic hydrocarbon Major group of organic compounds with a straight or branched molecular chain structure of carbon atoms. There are three groups: alkanes, alkenes, and alkynes.
Alkali A compound that has the ability to neutralize an acid and form a salt. *Example*: Sodium hydroxide referred to as caustic soda or lye. Alkali are more or less irritating, or caustic to the skin. Strong alkalies in solution are corrosive to the skin and mucous membranes.
Alkanes Saturated hydrocarbons—those containing only single bonds and relatively unreactive; paraffins.
Alkenes Unsaturated hydrocarbons—those containing one or more double bonds; olefins.
Alkynes Unsaturated hydrocarbons—those containing one or more triple bonds; acetylenes.
Alpha particles Particulate ionizing radiation consisting of helium nuclei traveling at high speed. Alpha particles have little penetrating energy and can be stopped by paper.
Alpha radiation A type of ionizing radiation consisting of alpha particles, which are two protons and two neutrons bound together, with an electrical charge of +2. An alpha particle is equivalent to a helium nucleus.
ALR Airline respirator.
Anhydrous Free from water.
Anion An ion with a net negative charge.
ANSI American National Standards Institute.
Applicable or relevant and appropriate requirements (ARARs) "Applicable" requirements are those cleanup standards, standards of control, and other environmental protection requirements promulgated under federal or state law that specifically address a hazardous substance, remedial action, location, or other circumstance at Superfund or CERCLA sites. "Relevant and appropriate" requirements are those cleanup standards which, while not applicable at a CERCLA site, address problems or situations similar to those encountered at CERCLA sites. ARARs can be action-specific, location-specific, or chemical-specific.
Aromatic hydrocarbons A major group of unsaturated cyclic hydrocarbons typified by benzene, which has a six-carbon ring with three resonating double bonds.

Asbestos Any material containing more than 1% asbestos in any form.

Asbestosis A disease of the lungs caused by the inhalation of fine airborne fibers of asbestos.

Asphyxiant A vapor or gas which can cause unconsciousness or death by suffocation. Most simple asphyxiants are harmful to the body only when they become so concentrated that they reduce oxygen in the air to dangerous levels (18% or lower). Asphyxiation is one of the principal potential hazards of working in confined places.

ASTM American Society for Testing and Materials

Atom The basic unit of an element that can enter into a chemical combination.

Atomic mass The mass of an atom in atomic mass units.

Atomic mass unit (amu) A mass exactly equal to 1/12th the mass of one carbon-12 atom.

Atomic number (Z) The number of protons in the nucleus of an atom.

Auto-ignition temperature The lowest temperature at which a flammable gas– or vapor–air mixture will ignite without a spark or flame being present. Along with the flashpoint, auto-ignition temperature gives an indication of relative flammability.

Avogadro's number 6.022×10^{23}; the number of particles in a mole.

Background radiation Radiation coming from sources other than the one directly under consideration. Background radiation due to cosmic rays and natural radioactivity is always present.

Base A substance that yields hydroxyl (OH^-) ions when dissolved in water. Bases react with acids form a salt. Base is another term for alkali.

Beta particles Negatively charged, particulate ionizing radiation consisting of electrons or positrons traveling at high speed.

Bioassay A technique by which a toxic agent is detected and measured for potency. The technique involves testing of the toxicant at different dosage levels for ability to cause a physiological response in a test organism.

Biocide A substance that, when absorbed by eating, drinking, or breathing, or otherwise consumed in relatively small quantities, causes illness or death.

Biohazard A combination of the words biological hazard. Infectious agents or products of infectious agents presenting a risk or potential risk to humans.

Biological half-life The time required to eliminate half of a substance which it takes in.

Biological hazardous wastes (infectious) Any substance of human or animal origin, other than food wastes, which are to be disposed of and could harbor or transmit pathogenic organisms including, but not limited to, pathological specimens such as tissues, blood elements, excreta, secretions, and related substances. This category included wastes from health care facilities and laboratories. Wastes from hospitals would include the following: malignant or benign tissues taken during autopsies, biopsies, or surgery; hypodermic needles; and bandaging materials.

Boiling point (B.P.) The temperature at which a liquid changes to a vapor state at a given pressure, usually expressed in degrees Fahrenheit at sea level. Flammable materials with low boiling points generally present special fire hazards.

Breakthrough time The elapsed time between initial contact of the hazardous chemical with the outside surface of a protective clothing material and the time at which the chemical can be detected at the inside surface of the material by means of the chosen analytical technique.

Breathing zone sample An air sample collected in the breathing area (around the head) of a worker to assess his exposure to airborne contaminants.

Bulk container A cargo container, such as that attached to a tank truck or tank car, used for transporting substances in large quantities.

Bulking Mixing together of chemicals in large quantities for transport.

Bung A cap or screw used to cover the small opening in the top of a metal drum or barrel.

°C Degrees centigrade (Celsius).

C or ceiling The maximum allowable human exposure limit for an airborne substance, not to be exceeded even momentarily. (See also PEL and TLV.)

CAA Clean Air Act.

Cancer A cellular tumor, the natural course of which is fatal and usually associated with formation of secondary tumors.

Cancer risk Incremental probability of an individual's developing cancer over a lifetime as a result of exposure to a potential carcinogen.

Canister (air-purifying) A container filled with filters, sorbents, and/or catalysts that remove particles, gases, and vapors from air drawn through the unit.

Carboy A container for holding liquids, with a capacity of approximately 5 to 15 gallons, made of glass, plastic, or metal, and often cushioned in a protective container.

Carcinogen A substance capable of causing cancer.

Cartridge A purifying device for an air-purifying respirator that attaches directly to the facepiece and removes particulates or specific chemical gases or vapors from the ambient air as it is inhaled through the cartridge.

CAS number Identifies a particular chemical by the Chemical Abstract Service number. This service indexes and compiles abstracts of worldwide information on chemicals.

Cation An ion with a net positive charge.

Caustic Something that strongly irritates, burns, corrodes, or destroys living tissue.

cc Cubic centimeter. A volume measurement in the metric system equal in capacity to one milliliter (ml). One quart is about 946 cubic centimeters.

Centigrade (Celsius) The internationally used scale for measuring temperature, in which 100 is the boiling point of water at sea level (1 atmosphere) and 0 is its freezing point.

CERCLA Comprehensive Environmental Response, Compensation and Liability Act (1980). Superfund.

CFR Code of Federal Regulations.

CGI Combustible gas indicator.

Chemical formula An expression showing the chemical composition of a compound in terms of the symbols for the atoms of the elements involved.

Chemical property Any property of a substance that cannot be studied without converting the substance into some other substance.

Chemical-resistant materials Materials that inhibit or protect against penetration of certain chemicals.

CHEMTREC Chemical Transportation Emergency Center, operated by the Chemical Manufacturers Association (CMA).

CHRIS Chemical Hazards Response Information System published by the United States Coast Guard.

Chronic effect Adverse effects resulting from repeated doses of, or exposures to, a substance over a relatively prolonged period of time.

Chronic exposure Exposure to a substance over a long period of time, usually at low doses.

Closed-circuit SCBA A type of self-contained breathing apparatus that recycles exhaled air by removing CO_2 and replenishing O_2. Also called a rebreather SCBA.

CNS Central nervous system.

Combustible Capable of burning.

Combustible liquid (DOT usage) Flashpoint 100°F to 200°F.

Compressed gas Material packaged in a cylinder, tank, or aerosol under pressure exceeding 40 psi at 70°F or other pressure parameters identified by DOT.

Concentration The relative amount of a substance when combined or mixed with other substances. Common methods of stating concentration are percent by weight or by volume, weight by unit volume, molarity, and so on. *Examples*: 2 ppm hydrogen sulfide in air or a 50% caustic solution.

Confined space An enclosure that is difficult to get out of and has limited or no ventilation. Examples are trenches, storage tanks, boilers, sewers, and tank cars.

Container Any portable device in which a material is stored, transported, disposed of, or otherwise handled.

Contamination control line The boundary between the support zone and the contamination reduction zone.

Contamination Reduction Corridor The part of the contamination reduction zone where the personnel decontamination stations are located.

Contamination Reduction Zone (CRZ) The area on a site where decontamination takes place, preventing cross-contamination from contaminated areas to clean areas.

Continuous flow respirator A respiratory protection device that maintains a constant flow of air into the facepiece at all times. Air flow is independent of user respiration.

Corrosive acid A liquid or solid, excluding poisons, that causes visible destruction or irreversible alterations in human skin tissue at the site of contact, or has a severe corrosion rate on steel. Liquids show a pH of 6.0 or less.

Corrosive alkaline A liquid or solid, excluding poisons, that causes visible destruction or irreversible alteration in human skin tissue at the site of contact, or has a severe corrosion rate on steel. Liquids show a pH of 8.0 or above.

Covalent bond A bond in which two electrons are shared by two atoms.

Covalent compounds Compounds containing only covalent bonds.

Crazing The formation of minute cracks (as in the lens of a facepiece).

CRC Contamination reduction corridor.

Cross-contamination The transfer of a chemical contaminant from one person, piece of equipment, or area to another that was previously not contaminated with that substance.

CRZ Contamination reduction zone.

CWA Clean Water Act, Title 40 CFR.

Dangerous when wet A label required for certain materials being shipped under US DOT regulations. Any of this labeled material that is in contact with water or moisture may produce flammable gases. In some cases, these gases are capable of spontaneous combustion.

Data evaluation This phase in the risk assessment process determines whether the available scientific data describe a causal relationship between an environmental agent and demonstrated injury to human health or the environment.

Decibel (dB) A unit used to express sound power level.

Decontamination The process of removing or neutralizing hazardous substances on personnel and equipment.

Decontamination line A specific sequence of decontamination stations within the contamination reduction zone for decontaminating personnel or equipment.

Degradation A chemical reaction between chemical and structural materials (in, for example, protective clothing or equipment) that results in damage to the structural material.

Demand respirator A respiratory protection device that supplies air or oxygen to the user in response to negative pressure created by inhalation.

Dermal Pertaining to skin.

Dermal toxicity Adverse effects resulting from skin exposure to a substance.

Dermatitis Inflammation of the skin from any cause.

Desiccant A substance such as silica gel that removes moisture (water vapor) from the air and is used to maintain a dry atmosphere in containers of food or chemical packaging.

Diatomic molecule A molecule that consists of two atoms.

Dilution A procedure for preparing a less concentrated solution from a more concentrated solution.

Disinfection The application of a chemical that kills bacteria and other microorganisms.

Displacement reaction An atom or an ion in a compound is replaced by an atom of another element.

Disposal drum/Salvage drum A nonprofessional reference to a drum used to overpack damaged or leaking containers of hazardous materials for shipment.

DOC Department of Commerce.
DOD Department of Defense.
DOE Department of Energy.
DOJ Department of Justice.
DOL Department of Labor.
DOS Department of State.
DOT Department of Transportation.
Dose The amount of energy or substance absorbed in a unit volume or organ or individual. Dose rate is the dose delivered per unit of time. (See also Roentgen, RAD, REM.)
Dose equivalent (DE) The product of absorbed dose, quality factor, and other modifying factors necessary to express on a common scale, for all ionizing radiations.
Dosimeter An instrument for measuring doses of radioactivity or other chemical exposures based on collection media.
dps Disintegrations per second. A unit of measure relating to the breakdown of a radioactive material.
Dressout area A section of the support zone where personnel suit up for entry into the exclusion zone.
Dust Solid particles generated by handling, crushing, grinding, rapid impact, detonation, and decrepitation of organic or inorganic materials, such as rock, ore, metal, coal, wood, and grain. Dusts do not tend to flocculate except under electrostatic forces; they do not diffuse in air but settle under the influence of gravity.
Edema A swelling of body tissues as a result of fluid retention.
Electrolyte A substance that, when dissolved in water, results in a solution that can conduct electricity.
Electron A subatomic particle that has a very low mass and carries a single negative electric charge.
Element A substance that cannot be separated into simpler substances by chemical means.
EPA United States Environmental Protection Agency.
Epidemiology The science that deals with the study of disease in a general population. Determination of the incidence (rate of occurrence) and distribution of a particular disease (as by age, sex, or occupation) may provide information about the cause of the disease.
Etiological agent A viable microorganism or its toxin, which causes or may cause human disease; limited to the agents identified in Title 42 CFR part 72.

Etiology The study of the causes of disease.

Evaporation rate The rate at which a particular material will vaporize (evaporate) when compared with the rate of vaporization of a known material. The evaporation rate can be useful in evaluating fire hazards of a material. The known material is usually normal butyl acetate (NBUAC or n-BuAc), with a vaporization rate designated as 1.0.

Exclusion zone Also known as the hot zone. The contaminated area of a site.

Exotoxin A toxin produced and delivered by a microorganism into the surrounding medium.

Explosive A chemical that is capable of burning or bursting suddenly and violently.

Explosive limits Some items have a minimum and maximum concentration in air which can be detonated by spark, shock, fire, and so on. The lowest concentration is known as the lower explosive limit (LEL). The highest concentration is known as the upper explosive limit (UEL).

Exposure Subjection of a person to a toxic substance or harmful physical agent in the course of employment through any route of entry (e.g., inhalation, ingestion, skin contact, or absorption); includes past exposure and potential (e.g., accidental or possible) exposure. An exposure to a substance or agent may or may not be an actual health hazard to the worker. An industrial hygienist evaluates exposures and determines if permissible exposure levels are exceeded.

Exposure assessment The study of identifying and characterizing exposure in potentially exposed populations.

The exposure analysis relies on many very different kinds of information, some based on actual measurements and some developed using mathematical models. The exposure analysis also develops "lifestyle" data to identify and describe populations likely to contact a pollutant.

If the estimated exposure for an environmentally exposed population is significantly smaller than the lowest dose producing a response in the study population, then the likelihood of injury to exposed humans is smaller. However, if the estimated exposure is significantly greater than the lowest dose, then the likelihood of injury is greater.

Exposure pathway The course a chemical or physical agent takes from a source to an exposed organism.

Exposure point A location of potential contact between an organism and a chemical or physical agent.

Exposure rate (absorbed dose rate) The time rate at which an exposure or absorbed dose occurs—that is, exposure or absorbed dose per unit time. It implies a uniform or short-term average rate, unless expressly qualified (e.g., peak dose rate). In protection work it is usually expressed in mR/hr, mrads/hr.

Exposure route The way a chemical or physical agent comes in contact with an organism (i.e., ingestion, inhalation, dermal contact).

°F Degrees Fahrenheit

Fahrenheit The scale of temperature in which 212 is boiling water at 760 mm Hg and 32 is the freezing point.

FFDCA Federal Food, Drug, and Cosmetic Act. (See Title 21 USC 301-392.)

Fibrosis A condition marked by increase of interstitial fibrous tissue.

FID Flame ionization detector.

FIFRA Federal Insecticide, Fungicide, and Rodenticide Act. (See Title 40 CFR.)

Film badge A pack of appropriate photographic film and filters used to determine radiation exposure.

Filter A purifying device for an air-purifying respirator that removes particulates and/or metal fumes from the ambient air as it is inhaled.

Flammable Capable of being easily ignited and/or burning with extreme rapidity.

Flammable (DOT usage) Flashpoint <100°F.

Flammable aerosol An aerosol which is required to be labeled "Flammable" under the United States Federal Hazardous Substances Labeling Act. For storage purposes, flammable aerosols are treated as Class 1A liquids (NFPA 30, Flammable and Combustible Liquids Code).

Flammable gas Any compressed or liquefied gas, except an aerosol, is flammable if either a mixture of 13% or less (by volume) with air forms a flammable mixture or the flammable range with air is wider than 12% regardless of the lower limit (at normal temperature and pressure).

Flammable limits Flammable limits produce (by evaporation) a minimum and maximum concentration of flammable gases in air that will support combustion. The lowest concentration is known as the lower flammable limit (LFL). The highest concentration is known as the upper flammable limit (UFL).

Flammable liquid Any liquid having a flashpoint below 100°F as determined by tests listed in 49 CFR 173.115(d). A *pyrophoric liquid* ignites spontaneously in dry or moist air at or below 130°F.

Flammable liquid class 1A (OSHA) Any liquid having a flashpoint below 73°F (22.8°C) and having a boiling point below 100°F (37.8°C) except any mixture having components with flashpoints of 100°F (37.8°C) or higher, the total of which comprise 99% or more of the total volume of the mixture (Title 29 CFR 1910.106).

Flammable solid (DOT usage) Any solid material, other than one classed as an explosive, that under conditions normally incident to storage is liable to cause fire through friction or retained heat from manufacturing or processing; or that can be ignited readily and, when ignited, burns so vigorously and persistently as to create serious storage hazard. Flammable solids, excluding Dangerous when wet, are further defined in Title 49 CFR 173.150.

Flashpoint The lowest temperature at which a liquid gives off enough vapor to form an ignitable mixture with air and produce a flame when a source of ignition is present. Two tests are used: open cup and closed cup.

Fp or fl. pt. Flashpoint.

ft^3 Cubic feet. Calculated by multiplying length by width by depth of an item or space.

Fully encapsulating suits Full chemical protective suits that are impervious to chemicals, offer full-body protection from chemicals and their vapors/fumes, and are to be used with self-contained breathing apparatus (SCBA).

Fume Gaslike emanation containing minute solid particles arising from the heating of a solid body such as lead. This physical change is often accompanied by a chemical reaction, such as oxidation. Fumes flocculate and sometimes coalesce. Odorous gases and vapors should not be called fumes.

FWPCA Federal Water Pollution Control Act (1972).

Gamma radiation, gamma rays Electromagnetic radiation of short wavelength and correspondingly high frequency, emitted by nuclei in the course of radioactive decay.

Gas A state of matter in which the material has very low density and viscosity; can expand and contract greatly in response to changes in temperature and pressure; easily diffuses into other gases; readily and uniformly distributes itself throughout any container. A gas can be changed to the liquid or solid state by the combined effect of increased pressure and/or decreased temperature.

GC/MC Gas chromatography/mass spectrometry. Refers to both analytical method and apparatus used for organic analysis.

Genetic effects Mutations or other changes which are produced by irradiation of the germ plasm.

g/kg Grams per kilogram, an expression of dose used in oral and dermal toxicity testing to indicate the grams of substance dosed per kilogram of animal body weight. (See also kg.)

GM Geiger–Mueller.

Grappler An implement used to hold and manipulate objects from a distance.

Half-reaction A reaction that explicitly shows electrons involved in either oxidation or reduction.

Halogens The nonmetallic elements in Group 7A of the Periodic Table (F, Cl, Br, I, an At).

Hazard assessment risk analysis A process used to qualitatively or quantitatively assess risk factors to determine mitigating actions.

Hazard class A category of hazard associated with hazardous materials or wastes that has been determined capable of posing an unreasonable risk to health, safety, and property when transported. (See Title 49 CFR 171.18.)

Hazardous air pollutant A pollutant to which no ambient air quality standard is applicable and that may cause or contribute to an increase in mortality or serious illness. For example, asbestos, beryllium, and mercury have been declared hazardous air pollutants.

Hazardous chemicals Chemicals or materials used in the workplace that are regulated under the OSHA Hazard Communication Standard or the "right-to-know" regulations in Title 29 CFR 1910.1200.

Hazardous material Any substance or mixture of substances having properties capable of producing adverse effects on the health and safety or the environment of a human being. Legal definitions are found in individual regulations.

Hazardous substances Chemicals, mixtures of chemicals, or materials subject to the regulations contained in Title 40 CFR.

Hazardous waste Any materials listed as such in Title 40 CFR 261, Subpart D, that possesses any of the hazard characteristics of corrosivity, ignitability, reactivity, or toxicity as defined in Title 40 CFR 261, Subpart C, or that is contaminated by or mixed with any of the previously mentioned materials. (See Title 40 CFR 261.3.)

Hazardous waste landfill An excavated or engineered area on which hazardous waste is deposited and covered; proper protection of the environmental from the materials to be deposited in such a landfill requires careful site selection, good design, proper operation, leachate collection and treatment, and thorough final closure.

Hazardous waste leachate The liquid that has percolated through or drained from hazardous waste stored in or on the ground.

Hazardous waste management Systematic control of the collection, source separation, storage, transportation, processing, treatment, recovery, and disposal of hazardous waste.

Hazardous waste manifest, uniform (EPA) The shipping document, originated and signed by the waste generator or his authorized representative, that contains the information required by Title 40 CFR 262, Subpart B.

Hazardous waste number The number assigned to each hazardous waste listed by EPA and to each hazardous waste characteristic.

Hazardous waste site A location where hazardous wastes are stored, treated, incinerated, or otherwise disposed of.

Hepatitis Inflammation of the liver.

Herbicide A chemical intended for killing plants or interrupting their normal growth.

HMTA Hazardous Materials Transportation Act (1975).

HNU Name of company that manufactures a type of photoionizer used to detect organic gases and vapors.

Hotline The outer boundary of the exclusion zone on a site.

HSWA Hazardous and Solid Waste Amendments of 1984.

Hydrocarbons Compounds made up only of carbon and hydrogen.

Hygroscopic Descriptive of a substance that has the property of absorbing moisture from the air, such as silica gel, calcium chloride, or zinc chloride.

Hypothermia Condition of reduced body temperature.

IDLH Immediately dangerous to life and health. An environmental condition which would immediately place a worker in jeopardy. Usually used to describe a condition existing where self-contained breathing apparatus must be used.

ID number Four-digit number by UN or NA, assigned to hazardous materials and dangerous goods.

Ignitable (EPA) A liquid with a flashpoint less than 140%.

Impoundment See Surface impoundment.

Incineration An engineered process using controlled flame combustion to thermally degrade waste materials. Devices normally used for incineration include rotary kilns, fluidized beds, and liquid injectors. Incineration is used particularly for the destruction of organic wastes with a high BTU value.

The wastes are detoxified by oxidations, and if the heat produced is high enough, they can sustain their own combustion and will not require additional fuel.

Incompatible Incapable of being combined without a dangerous effect—for example, descriptive of two or more substances that produce an unfavorable chemical reaction if they come in contact.

Industrial wastes Unwanted materials produced in or eliminated from an industrial operation. They may be categorized under a variety of headings, such as liquid wastes, sludge wastes, and solid wastes. Hazardous wastes contain substances that, in low concentrations are dangerous to life (especially human) for reasons of toxicity, corrosiveness, mutagenicity, and flammability.

Infectious wastes Waste that contains pathogens or consists of tissues, organs, body parts, blood, and body fluids that are removed during surgery or other procedures.

Ingestion The process of taking substances into the body, as in food, drink, medicine, and so on.

Inhalation The breathing in of a substance in the form of gas, vapor, fume, mist, or dust.

Injection To force, inject, or place a fluid or solid substance into the body.

Inner liner A continuous layer or lining of material placed inside a tank or other container that protects the construction materials of the tank or container from the contents.

Inorganic compounds Chemical compounds that do not contain the element carbon.

Inorganic matter Chemical substances of mineral origin, not containing carbon-to-carbon bonding. Generally structured through ionic bonding.

Insecticide A chemical product used to kill and control nuisance insect species.

Ion A charged particle formed when a neutral atom or group of atoms gain or lose one or more electrons.

Ionic bond The electrostatic force that holds ions together in an ionic compound.

Ionic compound Any neutral or uncharged compound containing anions and cations.

Ionization potential The energy required to remove the outermost electron from the molecule. The IP is specific for each compound and is measured in electron volts (eV).

Ionizing radiation High-energy radiation that causes irradiated substances to form ions, which are electrically charged particles.

Irritant Any material that produces an irritating effect upon contact with skin, eyes, nose, or respiratory system.

kg Kilogram. A metric unit of weight, about 2.2 United States pounds.

Label (DOT) Diamond-shaped, square-shaped, or rectangular-shaped attachment to a package or container that identifies the hazardous nature of a material. (See Title 49 CFR Part 172, Subpart E.)

Latent period The time which elapses between exposure and the first manifestation of damage or illness.

LC_{50} Lethal concentration 50, the concentration of a material which on the basis of laboratory tests is expected to kill 50% of a group of test animals when administered as a single exposure (usually 1 or 4 hours). Also, other LC values can be expressed (e.g., LC_{10} and LC_{20}).

LC_{Lo} Lethal concentration low. The lowest concentration of a substance in air, other than LC_{50}, which has been reported to have caused death in humans or animals. The reported concentrations may be entered for periods of exposure that are less than 24 hours (acute) or greater than 24 hours (subacute and chronic).

LD_{50} Median lethal dose. The dose which is required to produce death in 50% of exposed species. Death is usually reckoned as occurring within the first 30 days.

LD_{Lo} Lethal dose low. The lowest dose of a substance introduced by any route, other than inhalation, over any given period of time in one or more divided portions and reported to have caused death in humans or animals.

LEL Lower explosive limit. The lowest concentration of the material in air that can be detonated by spark, shock, fire, and so on.

LFL Lower flammable limit. The lowest concentration of the material in air that will support combustion from a spark or flame.

m^3 Cubic meter. A metric measure or volume, about 35.3 cubic feet or 1.3 cubic yards.

Manifest A list of cargo.

Manifest, uniform hazardous waste When properly prepared and distributed, provides a tracking system that consists of forms originating with the generator or shipper and following from the generator to disposal in a permitted TSDF.

Mass A measure of the quantity of matter contained in an object.

Mass number The total number of neutrons and protons present in the nucleus of an atom.
Matter Anything that occupies space and possesses mass.
Material safety data sheet (OSHA) See MSDS.
Melting point The temperature at which a material changes from a solid to a liquid.
Metals Elements that are good conductors of heat and electricity and have a tendency to form positive ions (cations) in ionic compounds.
Metalloid An element with properties intermediate between those of metals and nonmetals.
mg Milligram. A metric unit of weight. There are 1000 milligrams in one gram (g) of a substance.
mg/m³ Milligrams per cubic meter. A unit for measuring concentrations of dusts, gases, or mists in air.
Microorganism A living organism not discretely visible to the unaided eye. Bacteria, fungi, and so on.
Mineral A naturally occurring substance with a characteristic range of chemical composition.
Miscible Two liquids that are completely soluble in each other in all proportions are said to be miscible.
Mixture A combination of two or more substances in which the substances retain their identity.
ml Milliliter. A metric unit of capacity, equal in volume to one cubic centimeter (cc), or about 1/16 of a cubic inch. There are 1000 milliliters in one liter (liter).
mm Millimeter. A metric unit of length, equal to 1/1000 of a meter, or about 1/25 of an inch.
Molarity The number of moles of solute in one liter of solution.
Molecule An aggregate of at least two atoms in a definite arrangement held together by special forces.
MSDS Material safety data sheet. An MSDS must be in English and include information regarding the specific identity of hazardous chemicals. Also includes information on health effects, first aid, chemical and physical properties, and emergency phone numbers.
MSHA Mine Safety and Health Administration of the United States Department of Interior.
Mutagen A substance capable of causing genetic damage.
NA number North American identification number. When NA precedes a four-digit number, it indicates that this identification number is used in the

United States and Canada to identify a hazardous material or a group of hazardous materials in transportation.

NEPA National Environmental Policy Act (1969).

Neutralization The process by which acid or alkaline properties of a solution are altered by addition of certain reagents to bring the hydrogen ion and hydroxyl ion concentrations to an equal value; sometimes referred to as pH 7, the value of pure water.

Neutralize To make harmless anything contaminated with a chemical agent. More generally, to neutralize an acid or base by changing its pH to 7.

Neutron A subatomic particle that bears no net electric charge. Its mass is slightly greater than that of the proton.

NFPA National Fire Protection Association. An international voluntary membership organization to promote improved fire protection and prevention and establish safeguards against loss of life and property by fire.

NIOSH National Institute for Occupational Safety and Health. Federal agency which tests and certifies respiratory protective devices and air sampling detector tubes, recommends occupational exposure limits for various substances, and assists OSHA and MSHA in occupational safety and health investigations and research.

Nonelectrolyte A substance that, when dissolved in water, gives a solution that is not electrically conducting.

Nonmetals Elements that are usually poor conductors of heat and electricity.

Nonflammable gas Any material or mixture, in a cylinder or tank, other than poison gas or flammable gas having in the container an absolute pressure exceeding 40 psi at 70°F, or having an absolute pressure exceeding 104 psi at 130°F (Title 49 CFR and CGA).

NPDES National Pollutant Discharge Elimination System (water quality usage).

NRC National Response Center (800-424-8802) or Nuclear Regulatory Commission (10 CFR usage).

Nucleus The central core of the atom.

Nuisance material Materials causing localized irritation to the eyes, skin, mucous membranes, and respiratory tract. They cause no long-term or systemic effects.

Olfactory Relating to the sense of smell.

Open-circuit SCBA A self-contained breathing apparatus in which the user exhales air directly to the atmosphere.

Oral toxicity Adverse effects resulting from taking a substance into the body through the mouth.

Organic compounds Compounds that contain carbon, usually in combination with elements such as hydrogen, oxygen, nitrogen, and sulfur.

Organic peroxide Any organic compound containing the bivalent —O—O— structure and that may be considered a derivative of hydrogen peroxide where one or more of the hydrogen atoms have been replaced by organic radicals. Peroxides can initiate or promote combustion in other materials, thereby causing fire through the release of oxygen or other gases.

OSHA Occupational Safety and Health Administration of the United States Department of Labor. Federal (or State) agency with safety and health regulatory and enforcement authorities for most United States industry and business.

OVA Organic vapor analyzer.

Overpack Except when referring to a packaging specified in Title 49 CFR Part 178, this an enclosure used by a single consignor to provide protection or to consolidate two or more packages. Overpack does not include a freight container.

Oxidation number The number of charges (positive or negative) an atom has in a molecule due to the transfer of electrons.

Oxidation reaction The half-reaction that involves the loss of electrons.

Oxidation state See oxidation number.

Oxidizer A chemical other than a blasting agent or explosive as defined in Title 29 CFR 1910.109(a) that initiates or promotes combustion in other materials, thereby causing fire either of itself or through the release of oxygen or other gases.

Oxidizing Agent A substance that can accept electrons from another substance or increase the oxidation number in another substance.

Palletize To place on a pallet; to transport or store by means of a pallet.

Pallets A low portable platform constructed of wood, metal, plastic, or fiberboard, built to specified dimensions, on which supplies are loaded, transported, or stored in units.

Particulate Formed of separate, small, solid pieces.

Pathogen Any microorganism capable of causing disease.

PCB Polychlorinated biphenyl (see Title 40 CFR 761.3).

PCB-contaminated electrical equipment Any electrical equipment, including transformers that contain at least 50 ppm but less than 500 ppm PCB (Title 40 CFR 761.3).

PCB transformer Any transformer that contains 500 ppm PCB or greater (Title 40 CFR 761.3).

PCP (a) Abbreviation for pentachlorophenol (q.v.), a wood preservative.

PEL Permissible exposure limit. An exposure limit established by OSHA regulatory authority. May be a time-weighted average (TWA) limit or a maximum concentration exposure limit. (See also Skin.)

Permeation Seepage and sorption of a chemical through a material (e.g., the material making up protective clothing or equipment).

Pesticide General term for that group of chemicals used to control or kill such pests as rats, insects, fungi, bacteria, weeds, and so on. Among these are insecticides, herbicides, fungicides, rodenticides, miticides, fumigants, and repellents.

PF Protective factor. Refers to the level of protection a respiratory protective device offers. The PF is the ratio of the contaminant concentration outside the respirator to that inside the respirator.

pH A measure of hydrogen ion concentration [H^+]. A pH of 7.0 is neutrality; higher values indicate alkalinity and lower values indicate acidity.

Physical property Any property of a substance that can be observed without transforming the substance into some other substance.

PID Photoionization detector.

Poison Any substance which when taken into the body is injurious to health.

Pollution Contamination of air, water, land, or other natural resources that will, or is likely to, create a public nuisance or render such air, water, land, or other natural resources harmful, detrimental, or injurious to public health, safety, or welfare.

Polychlorinated biphenyl (PCB) Any of 209 compounds or isomers of the biphenyl molecule that have been chlorinated to various degrees (see PCB).

POTW Public-owned treatment works.

ppb Parts per billion. A unit for measuring the concentration of a gas or vapor in air; parts (by volume) of the gas or vapor in a billion parts of air. Usually used to express measurements of extremely low concentrations of unusually toxic gases or vapors. Also used to indicate the concentration of a particular substance in a liquid or solid.

PPE Personal protective equipment.

ppm Parts per million. A unit for measuring the concentration of a gas or vapor in air; parts (by volume) of the gas or vapor in a million parts of air. Usually used to express measurements of extremely low concentrations of unusually toxic gases or vapors. Also used to indicate the concentration of a particular substance in a liquid or solid.

Pressure-Demand Respirator A respiratory protection device that supplies air to the user and maintains a slight positive pressure in the facepiece at all times and also supplies additional air in response to the negative pressure created by inhalation.

Pretreatment standards (CWA) Specific industrial operation or pollutant removal requirements in order to discharge to a municipal sewer.

Proper shipping name The name of the hazardous material shown in Roman print (not italics) in Title 49 CFR 172.101 or 172.102 (when authorized).

Protection factor The ratio of the ambient concentration of an airborne substance to the concentration of the substance inside the respirator at the breathing zone of the wearer. The protection factor is a measure of the degree of protection that the respirator offers.

Proton A subatomic particle having a single positive charge. The mass of the proton is about 1840 times that of the electron.

psi Pounds per square inch.

psia Pounds per square inch absolute.

psig Pounds per square inch gauge.

Pulmonary Pertaining to the lungs.

PVC Polyvinyl chloride.

Pyrophoric A chemical that will ignite spontaneously in air at a temperature of 130 degrees Fahrenheit (54.4 degrees Celsius) or below.

Rad A unit for the measurement of radioactivity. One rad is the amount of radiation that results in the absorption of 100 ergs of energy by 1 gram of material.

Radiation Energy in the form of electromagnetic waves.

Radioactive material An element or isotope that is characterized by its spontaneous decay and emission of ionizing radiation.

RCRA Resource Conservation and Recovery Act (1976).

Reagent A substance used in a chemical reaction to detect, measure, examine, or produce other substances.

Recovery drum A nonprofessional reference to a drum used to overpack damaged or leaking hazardous materials. (See Disposal drum.)

Redox reaction A reaction in which there is either a transfer of electrons or a change in the oxidation numbers of the substances taking part in the reaction.

Redress area A section of the exclusion zone where decontaminated personnel put on clothing for use in the support zone.

Reducing agent A substance that can donate electrons to another substance or decrease the oxidation numbers in another substance.

Reduction reaction The half-reaction that involves the gain of electrons.

Reference dose (RfD) A toxicity value for evaluating potential noncarcinogenic effects in humans resulting from contaminant exposures at CERCLA sites.

Rem A measure of radiation dose meaning roentgen equivalent man. The dose in rems is calculated by multiplying the dose in rads by the relative biological effectiveness of the radiation considered.

Reportable quantity (EPA) CERCLA established a list of hazardous materials and their reportable quantities (RQs). These materials when released into the environment in an amount equal to or greater than their RQs, in any 24-hour period, must be reported to the National Response Center and any local agencies requiring this information. The RQ for trichloroethylene is 100 lb. The RQ for vinyl chloride is 1 lb and the RQ for ethyl acetate is 5000 lb.

Respiratory system Consists of the nose, mouth, nasal passages, nasal pharynx, pharynx, larynx, trachea, bronchi, bronchioles, air sacs (alveoli) of the lungs, and muscles of respiration.

Risk assessment An investigation of the potential risk to human health or the environment posed by a specific action or substance. The assessment usually includes toxicity, concentration, form, mobility, and potential for exposure of the substance.

Risk characterization Risk characterization is the final step in the risk assessment process, where important information, data, and conclusions from the data evaluation, toxicity assessment, and exposure assessment are examined together to characterize risk. The risk characterization is the product of the risk assessment process and is much more than a number of value. While the risk is often stated as a bare number—for example, "a risk of 10^{-6}" or "one in a million new cancer cases"—the analysis involves substantially more information, thought, and judgment than the numbers express.

Roentgen A measure of the charge produced as the rays pass through air.

RQ See Reportable quantity.
Salt An ionic compound made up of a cation other than H^+ and an anion other than OH^- or O^{2-}.
Salvage drum A drum with a removable metal head that is compatible with the lading used to transport damaged or leaking hazardous materials for repackaging or disposal. Also referred to as disposal or recovery drum.
SARA Superfund Amendments and Reaurthorization Act (1986).
SCBA Self-contained breathing apparatus.
SDWA Safe Drinking Water Act (1974).
Sensitizer A substance which on first exposure causes little or no reaction in humans or test animals, but which on repeated exposure may cause a marked response not necessarily limited to the contact site. Skin sensitization is the most common form of sensitization in the industrial setting, although respiratory sensitization to a few chemicals is also known to occur.
"Skin" A notation, sometimes used with PEL or TLV exposure data; indicates that the stated substance may be absorbed by the skin, mucous membranes, and eyes—either airborne or by direct contact—and that this additional exposure must be considered part of the total exposure to avoid exceeding the PEL or TLV for that substance.
Smoke An air suspension (aerosol) of particles, often originating from combustion or sublimation. Carbon or soot particles less than 0.1 μm (micron) in size result from the incomplete combustion of carbonaceous materials such as coal or oil. Smoke generally contains droplets as well as dry particles.
Solubility The maximum amount of solute that can be dissolved in a given quantity of solvent at a specific temperature.
Solute The substance present in smaller amount in a solution.
Solution A homogeneous mixture of two or more substances.
Solvent The substance present in larger amount in a solution.
SOP Standard operating procedure.
Sorbent material A substance that takes up other materials by either absorption or adsorption. Sorbent materials are used for spill cleanup.
SPCC plan (CWA) Spill prevention, control, and countermeasure plan.
Staging area An area in which items are arranged in some order.
Standard operating procedure (SOPs) Established or prescribed tactical or administrative methods to be followed routinely for the performance of designated operations or in designated situations.
STEL Short-term exposure limit: ACGIH terminology.

Storage When used in connection with hazardous waste, it means the containment of hazardous waste, either on a temporary basis or for a period of years, in such a manner as not to constitute disposal of such hazardous waste.

Storage facility Any facility used for the retention of hazardous waste prior to shipment or usage, except generator facilities (under Title 40 CFR), which are used to store wastes for less than 90 days, for subsequent transport.

Storage tank Any manufactured, nonportable, covered device used for containing pumpable hazardous wastes.

Strict liability The defendant may be liable even though he may have exercised reasonable care.

Substance A form of matter that has a definite or constant composition and distinct properties.

Superfund A common name for the Comprehensive Environmental Response, Compensation, and Liability Act (CERCLA) of 1980.

Supplied air respirator A respiratory protection device that supplies air to the user from a source that is not worn by the user but is connected to the user by a hose. Also called an airline respirator.

Support zone The uncontaminated area of a site where workers will not be exposed to hazardous conditions. Also referred to as the clean zone. The offices, laboratories, and dressing areas are usually located in the support zone.

Surface impoundment Any natural depression or excavated and/or diked area built into or upon the land, which is fixed, uncovered, and lined with soil or a synthetic material and is used for treating, storing, or disposing wastes. Examples include holding ponds and aeration ponds.

Swab A piece of cotton or gauze on the end of a slender stick used for obtaining a piece of tissue or secretion for bacteriologic examination.

Swipe A patch of cloth or paper that is wiped over a surface and analyzed for the presence of a substance.

Synergism Cooperative action of substances whose total effect is greater than the sum of the their separate effects.

Teratogen A substance or agent which can result in malformations of a fetus.

Threshold The level where the first effects occur or are noticeable. The odor threshold is the lowest concentration at which a material can be sensed by its odor.

TLV Threshold limit value. An exposure level under which most people can work consistently for 8 hours a day (day after day) with no harmful effects. A table of these values and accompanying precautions is published annually by the American Conference of Governmental Industrial Hygienists (ACGIH).

TLV-C Threshold limit value—ceiling.

Toxicity A relative property of a chemical agent; refers to a harmful effect on some biological mechanism and the condition under which this effect occurs.

Toxicity assessment The dose–response analysis is designed to establish the quantitative relationship between exposure (or dose) and response in existing studies in which adverse health or environmental effects have been observed.

TSCA or TOSCA Toxic Substances Control Act (1976).

TSDF Treatment, storage, or disposal facility.

TWA Time-weighted average exposure. The airborne concentration of a material to which a person is exposed, averaged over the total exposure time—generally the total workday. (See also TLV.)

TWA-C Time weighted average—ceiling limit. The excursion limit placed on fast acting substances that limits all exposures below the applicable C limit. All time weighted average concentrations and peak exposures must be less than this limit.

UEL Upper explosive limit. The highest concentration of the material in air that can be detonated.

UFL Upper flammable limit. The highest concentration of the material in air that will support combustion.

UL Underwriters Laboratories, Inc.

UN number United Nations identification number. When UN precedes a four-digit number, it indicates this identification number is used internationally to identify a hazardous material.

USCG U.S. Coast Guard

UST Underground Storage Tanks (See LUST).

UV Ultraviolet.

Valence The number of electron involved in bond formation. Valence is also called oxidation number or oxidation state.

Vapor An air dispersion of molecules of a substance that is liquid or solid in its normal physical state, at standard temperature and pressure. Examples are water vapor and benzene vapor. Vapors or organic liquids are loosely called fumes; however, it is not technically appropriate to use the term fume for vapors of organic liquids.

Vapor density The ratio of the vapor weight of the commodity compared to that of air. Vapors will diffuse and mix with air due to natural air currents. In general, if the ratio is greater than 1, the vapors are heavier and may settle to the ground; if lower than 1, the vapors will rise.

Vapor pressure The pressure of the vapor in equilibrium with the liquid at the specified temperature. Higher values indicate higher volatility or evaporation rate. Dangerous, flammable materials have high vapor pressures.

X-radiation, X-rays Electromagnetic ionizing radiation originating outside the atomic nucleus. X-rays are indistinguishable from gamma rays of the same energy, the distinction being one's source.

INDEX

Accident investigation, 348
Accident reporting:
  regulatory requirements, 349
Acid rain, 43
Acidic solutions, 48
Acids, 50, 106
  and bases, 47
Acne, 78
Active samplers, pumps, 273
Active sampling systems, 274
Acute health effect, 70
Agranulocytosis, 103
Air monitoring, 187, 227, 281, 286, 335, 340, 347
  direct-reading instruments, 230
  instruments, 229
  oxygen meters, 232
  purposes, 228
Air monitoring strategies, 281
Airborne contaminants, 76
Airline respirators, 198
  constant-flow, 199
  demand-flow, 199, 222
  pressure-demand, 201
Air-purifying respirators, 194, 195

Aliphatic hydrocarbons, 58
Alkali metals, 32
Alkaline earth metals, 34
Alkalis, 106
Alkanes, 58
Alkenes, 59
Alkynes, 60
Allergic contact dermatitis, 77
Alpha particles, 121
American Conference of Governmental Industrial Hygienists (ACGIH), 84
Analysis of blood samples, 87
Analysis of exhaled air, 87
Analysis of urine, 87
Anesthetics and narcotics, 72
Animal studies, 82
Anions, 39, 47, 53
Annual exam:
  medical surveillance, 369
Anthracene, 63
Area samples, 283
Area sampling, 271
Aromatic hydrocarbons, 61
Arsenic, 53
Asbestos, 73

Asbestosis, 74, 104
Asphyxial anoxia, 102
Asphyxiants, 72, 76
Atom, 31
Atomic mass, 41
    atomic weight, 36, 42
Atomic number, 31
Avagadro's number, 42

Balancing chemical Reactions, *see* Chemical reactions
Barium, 53
Bases, 50
Basic solutions, 48
Benzene, 61
Benzene series hydrocarbons, 61
Berylliosis, 74
Beryllium, 53
Beta particles, 121
Biological exposure indices, 86
Black lung disease, 74
Blood and hematopoietic or blood-forming system, 103
Blood chemistry, 374
Boiling point, 90
Botulinum toxin, 102
Bromobenzene, 98
BTEX, 61
Bulging drums, 145
Butane, 59
Byssinosis, 104

Cadmium, 53, 80, 97, 98
Calibration:
    instruments, 236
    personal samplers, 279
Cancer:
    risk assessment, 391
Carbon atom:
    organic chemistry, 56
Carbon monoxide, 96
Carbon tetrachloride, 98
Carbon-12, 41
Carcinogens, 70, 94, 283
    risk assessment, 401
Care of respirators, 207
Cations, 35, 47

CERCLA, *see* Comprehensive Environmental Resource Compensation and Liability Act
Chemical analogy, 82
Chemical and physical hazards, 88
Chemical and physical properties of substances, 90
Chemical asphyxiants, 96
Chemical decontamination methods, 296
Chemical definitions, 30
Chemical property, 31
Chemical protective clothing (CPC), 164
Chemical reactions, 42
Chemical resistance, 166, *see* Chemical protective clothing
Chemicals at NPL sites, 66
Chemistry:
    definition, 29
Chilblains, 118
Chimney sweeps, 79
Chlorinated aromatic hydrocarbons, 63
Chlorinated benzenes, 64
Chloroform, 98
Chromium, 54, 98
Chromium compounds, 41
Chronic health effect, 70
Coal tar compounds, 107
Cold disorders, 117
Cold stress, 115
    prevention, 119
Colorimetric tubes, *see* Detector tubes
Combination airline respirators with egress bottle, 199
Combustible gas indicator and oxygen meter model, Model 261, 232
Combustible gas indicators (CGI), 232
Communication systems:
    work practices: 325
Compound, 35
Comprehensive Environmental Resource Compensation and Liability Act (CERCLA), 1, 16
Compressed gas, 89
Concentration measurements, 228
Conductive hearing loss, 112
Confined space:
    air monitoring, 333

definitions, 328
entry, 328
entry equipment, 332
permit system, 331
rescue, 338
Contact dermatitis, 77
Contamination reduction zone, *see* Decontamination zone
Corrosion, 90
of skin, 77
Corrosive, 95
material, 25
Corrosivity, 15
Covalent bonding, 36
CPC, inspection, maintenance, and storage, 179
Cyanide poisoning, 96
Cyclic aliphatic hydrocarbons, 60
Cyclopropane, 60
Cytotoxic anoxia, 103
Cytotoxic hypoxia-cyanide, 103
Cytoxic hypoxia, 96

Data evaluation:
risk assessment, 389, 396
Decomposition reactions, 43
Decon station design, 297
Decontamination, 291
effectiveness testing, 297
equipment, 298
of heavy equipment, 305
line, 303
line layouts
levels A, B, and C, 300, 301, 302, 303, 304
methods, 294
plan, 293
procedures, 298
of tools, 304
zone, 293
Degradation, 168
Department of Transportation (DOT), 20
system of placarding and labeling, 127
Department of Transportation Act of 1966, 20

Detector tubes (colorimetric indications), 260
qualitative exercise, 266
theory, application, and use, 261
Dichlorodiphenyl trichloroethane (DDT), 102
Diffusion samplers, 278
Dimethyl nitrosamine, 98
Direct chemical combination or synthesis, 42
Direct-reading colorimetric tubes, 276
Direct-reading instruments, 230
Dissociation, 47
Dose-response curve, 81
Dose-response relationship, 80
DOT definition, combustible liquid, 23, 89
of hazardous material, 20
DOT hazard classes, 21
DOT—HM-181, 20
Double bonds, 38
Double displacement, substitution or replacement reactions, 43
Drum handling, 137, 142, 145
Drum sampling, 148, 149, 150
Drum type, 136, 138, 143, 145
Drums, 135, 137, 139, 159, 160
Dusts, 76

Effects of noise, 11
Electrolytes, 47
Electrostatic attraction, 35
Electrostatic forces, 39
Element, 31
Emergency decontamination, 309
Emergency Response Guidebook, 129
Emergency response planning, 340
Employee:
access to medical records, 372
responsibilities, 9, 12
rights, 8, 9
Employer:
responsibilities, 9, 13
rights:
employee medical records, 373
Environmental Protection Agency (EPA), 1

Environmental Protection Agency (*Continued*)
  advisory protocol for hazardous waste site entry, 165, 167
  definition, ignitable liquid, 89
Epoxy resin hardeners, 106
Ethane, 58
Ethyl benzene, 61, 62
Examples of hazardous chemicals requiring medical surveillance, 376
Explosives, 22, 89
  (or flammable) range, 91, 234
  limits, 234
Exposure assessment:
  risk assessment, 390, 396
Eyes, 107
  exposure, 78

Fibrosis, 73
Field laboratory:
  support zone activities, 325
Fire, 108
  protection:
    fire extinguishers, 346
  pyramid, 109
  tetrahedron, 108
  triangle, 108
First aid equipment, 344
Flame ionization detectors (FID), 238, 242
Flammability, 91
Flammable liquids, 23
Flammable materials, 89
Flammable solids, 23
Flash point, 91
French Le Systemè International d'Unitès, 47
Frostbite, 118
Frostnip, 119
Fumes, 76

Gamma rays, 121
Gaseous contaminants, 187
Gases, 23, 76
General duty clause of the OSH Act, 13
General industry standards, 13

Generic respiratory protection program, 211

Half-reactions, 39
Halogenated hydrocarbons, 98
Halogenated pesticides, 97
Halogens, 34
Handling drums, 137, 142, 145
Hazard recognition, 125
Hazardous Material Information System (HMIS), 132, 134
Hazardous substance, 70
Hazardous waste, 70
HAZWOPER, 17
Health and safety hazards, 126
Hearing damage, 111
Hearing loss, 112
Heat:
  cramps, 114
  exhaustion, 114
  rash, 114
Heat stress, 114
  causes, 114
  prevention, 115
Heat stroke, 114
Heavy metals, 53
Hematotoxins, 96
Hemolytic anemia and hemolysis, 103
Hepatotoxins, 75, 95
Highly toxic agents, 94
Hot zone, 292, 293
Human epidemiological studies, 83
Hydrogen, 34
Hydrogen sulfide poisoning, 96
Hypothermia, 117
Hypoxia (anoxia) and chemical asphyxiants, 103

Identification of airborne contaminants, 283
Ignitability, 15
Impingers and bubblers, 275, 276
Industrial hygiene, 69
Infectious materials, 134
Infectious waste, 134
Ingestion of toxins, 78
Inhalation, 75, 186, 210

Initial site characterization and analysis, 314
Initial site entry:
  work practices, 315
Injection, 78
Intermolecular forces, 39
International System of Units, 41
Ionic bonding, 35
Ionization, 48
  of water, 48
  potential (IP), 240, 243
Ionizing radiation, 120
Irritants, 72, 95, 105
Irritant contact dermatitis, 105
Isomers, 57

Kidneys, 98

Labels, 127
Laboratory packs, 143, 144
Latency period, 94
LD_{100}, 80
Lead, 54, 80, 97, 98
Lethal concentration (LC), 80
Lethal concentration 50 (LC_{50}), 80
Lethal dose, 80
Lethal dose 50 (LD_{50}), 80
Levels of protection, 166, 171
Lewis dot symbols, 35
Liver, 98
Lower explosive limit (LEL), 91, 234
Lungs, 103

Mass, 30
Material safety data sheets (MSDSs), 151, 154
Matter, 30
Maximum use limitation, 188
Medical exam:
  notification, 371
  content, 371
Medical surveillance program, 369
Medical test and protocols, 374
Membrane filters, 277
Mercury, 54, 80, 98
Mesothelioma, 74
Metabolism of toxins, 79

Metallic mercury, 98
Methane, 58
Methyl mercury, 97
Micrograms per liter (μg/L), 46
Milligrams per liter (mg/L), 46
Mists, 76
Mixtures, 30
Molar solution, 46
Molarity, 46
Molecular formulas, 34
Molecular mass:
  molecular weight, 42
Molecules, 34

Naphthalene, 62
National Institute for Occupational Safety and Health (NIOSH), 8, 192
National Response Center, 16
Negative pressure test, 190
Nephrotoxic agents, 75, 96
Nephrotoxins, 98
Nervous system, 102
Neurotoxins, 71, 75, 96
Neutralization, 51
Neutrons, 31, 122
NFPA-704M diamond, 130, 133
Nickel, 54, 97
Nitrosamines, 98
Nitrous oxides, 104
Noise levels, 110
Noise, administrative or engineering controls, 112
  definition, 110
Noncarcinogenic effects:
  threshold concept, 400
Noncarcinogens:
  risk assessment, 399
Nonelectrolytes, 47
Notice of Intended Changes, 88
Nuisance materials, 71
Number placard, 128

Occupational acne, 107
Occupational asthma, 105
Occupational Safety and Health Act (OSHA), 7, 8, 13

438 ■ INDEX

Occupational Safety and Health Act
(*Continued*)
   hazard communication, 88, 152
   inspection, 9
   offices
      medical surveillance guidance, 384
Oils, 106
Olefins, 59
Organic chemistry, 55
   definition, 55
   properties, 55
Organic peroxides, 24, 28, 89
Organic solvents, 97, 106
Organophosphate insecticides, 79
Other regulated materials (ORM-D), 25
Oxidants, 107
Oxidation number, 40
Oxidation state, 40, 41
Oxidation–reduction reactions, 39
Oxidizers, 41
Oxidizing agent, 40, 89
Oxidizing materials, 24
Oxygen meters, 232

Polycyclic aromatic hydrocarbons (PAHs), 62
Paraffins, 58
Particulate collectors, 276
Particulate hazards, 210
Parts per million, 46
Passive samplers, 278
   diffusion, 272
Penetration, 166
Pentane, 59
Percent by weight, 45
Periodic table, 32, 33
Permanent threshold shift, 113
Permeation, 168, 178, 201
   influencing factors, 294
   samplers, 278
Permissible exposure limits (PELs), 85
Permissible levels of harmful noise, 112
Permit-required confined spaces
      procedures, 331
Personal air sampling, 271
Personal protective equipment (PPE), 163, 164, 165, 167

Personal sampling, 271
   plan, 279
   pumps, 273
pH, 48, 92
Phenanthrene, 63
Photochemical oxidants, 103
Photoionization detectors (PID), 238, 243, 245, 261
Physical decontamination methods, 294
Physical hazards, 107
Physical property, 31
Physical state, 92
Pigment disturbances, 107
Placards, 127
Planning for work site activities, 313
Poisons, 24
Polychlorinated biphenyls (PCBs), 64
Polycyclic aromatic hydrocarbons, 62
Positive pressure test, 190
Preplacement exam:
   medical surveillance, 370
Presbycusis, 113
Products of chemical reactions, 42
Propane, 58
Protection factors, 188
Pyrethrins (pesticides), 102
Pyrophoric materials, 24, 89

Qualitative fit-test, 191
Quantitative fit-test, 191, 216

Radiation, 120
Radioactive material, 25
Reactants, 42
Reactivity, 15
Real-time monitoring, 286
Record-keeping requirements, 372
Redox reaction, 40
Reducing agent, 40, 45
Relative response factors, *see*
      Photoionization detectors (PID)
Relative responses:
   combustible gas indicators, 236
   of the OVA, 263
Remedial alternatives:
   risk assessment, 404
Renal damage, 80

Reproductive system, 97
Reproductive toxins, 75, 96
Resource Conservation and Recovery
    Act (RCRA), 14
  heavy metals, 55
  major provisions of, 14
  definition of hazardous waste, 14
Respirator fit-testing, 189
Respiratory equipment, 192
Respiratory protection, 185
Results of risk characterization, 403
Risk assessment, 71
  disciplines involved, 388
Risk assessment in superfund, 387, 392
Risk assessment process:
  in superfund, 393
Risk characterization:
  risk assessment, 391, 402
Risk management, 388
Routes of toxicant entry, 75

Safety, 71
Safety equipment
  emergency showers, 346
  requirements
    eye wash, 345
Sample bags, 276
Sample collection devices, 294
Sampling methods, 271
Sampling ponds and lagoons, 149, 151, 152, 173
Sampling tube, 275
SCBA, use and service, 206
Scrotal cancer, 79
Selection criteria for CPC, 170
Selenium, 54
Self-contained breathing apparatus, 202
Sensitizer(s), 74, 85
Sensorineural hearing damage, 112
Silica, 73
Silicosis, 73, 104
Single bond, 38
Single displacement, substitution or replacement reactions, 43
Site control, 318
Site emergency procedures, 341
Site preparation:
  work practices, 317
Site safety plan, 353
  air monitoring requirements, 352
  regulatory requirements, 351
  requirements, 351
Site security, 328
Site work zones, 318
Skin, 105
  cancers, 78
  exposure, 77
  notation, 85
Soaps, 106
Solubility, 92
Solute, 45
Solution(s), 44, 45
Solution concentration, 45
Solvent(s), 45, 99, 107
Specific gravity, 92
Stimulants, 102
Strong acids, 51
Strong bases, 51
Styrene, 62
Substance, 30
Sulfur oxides, 103
Superfund, 16
Superfund Amendments and
    Reauthorization Act (SARA), 1, 17, 351
  Title I, 16
  Title II, 17
  Title III, 17
Supplied-air respirators, 198
Support zone, 293, 323
  activities, 324
Suspension, 45
Systemic, 76
Systemic poisons, 75

Table of Placards, 130
Target organs, 97
Target tissues, 79
Temperature stress, 113
Temporary threshold shift, 113
Teratogenic agents, 71
Termination exam:
  medical surveillance, 370
Tetrachloroethylene, 98

Tetrodotoxin, 102
Thallium, 55
Threshold limit value-ceiling (TLV-C), 85
Threshold limit value-short term
    exposure limit (TLV-STEL), 84
Threshold limit values (TLVs), 83
Threshold limit value-time-weighted
    average (TLV-TWA), 84
Thrombocytopenia, 103
Tinnitus, 112
Toluene, 61
Total organic vapor detectors, 238
Toxic agents, 95
Toxic atmospheres, 237
Toxic chemicals, 79
Toxic gas detectors, 237
Toxicant ingestion, 78
Toxicity, 71
    assessment
        risk assessment, 389, 399
    measurements, 80
Toxicology, 69
Trace inorganic contaminants, 53
Training:
    confined space entry, 328
Training requirements of HAZWOPER
    SARA, 17
Transportation Safety Act of 1974, 20

Trichloroethylene(1,1,2), 98
Triple bond, 38
Types of contamination, 293

Ulceration, 107
UN (United Nations) hazard class, 128
Unstable material, 90
Upper explosive limit (UEL), 234
Upper explosive or flammable limit, 91
Uranium, 102

Valence, 40
Vapor density, 93
Vapor pressure, 93
Vapors, 76
Viscosity, 94
Volatility, 94, 108

Waste site hazards, 2
Water reactive materials, 24, 90
Weak acids, 50, 51
Weak bases, 51
Weight, 30
Work practices, 313
Work zones:
    decontamination of, 291

X-rays, 121